T025729

D0995047

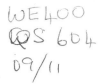

WE400
QS 604
09/11

# CRANIOFACIAL EMBRYOGENETICS AND DEVELOPMENT

## Second Edition

**Geoffrey H. Sperber**
BSc (Hons), BDS, MS, PhD, FICD
Dr Med Dent (Honoris Causa), For Assoc RSS Af
Professor Emeritus
Faculty of Medicine and Dentistry
University of Alberta
Edmonton, Alberta
Canada
T6G 2N8

**Steven M. Sperber**
BSc, MSc, PhD, FACMG
GeneDx
Gaithersburg, Maryland

**Geoffrey D. Guttmann**
PhD
Professor of Anatomy
The Commonwealth Medical College
Scranton, Pennsylvania

With a Foreword by:

**Phillip V. Tobias**
DSc, PhD, MBBCh (Witwatersrand)
FRS, FRCP, Hon FCMSA, Hon FRSSAf, NAS
ScD (Honoris Causa) (Cantab, Pennsylvania)
DSc (Honoria Causae (mult), Hon Dr Med et Chir (Turin)
Professor of Anatomy and Human Biology Emeritus
University of the Witwatersrand
Johannesburg, South Africa
Andrew Dickson White Professor-at-Large
Cornell University
Ithaca, New York

2010
PEOPLE'S MEDICAL PUBLISHING HOUSE—USA
SHELTON, CONNECTICUT

**People's Medical Publishing House–USA**
2 Enterprise Drive, Suite 509
Shelton, CT 06484
Tel: 203-402-0646
Fax: 203-402-0854
E-mail: info@pmph-usa.com

**PMPH-USA**

© 2010 Geoffrey H. Sperber.

All rights reserved. Without limiting the rights under copyright reserved above, no part of this publication may be reproduced, stored in or introduced into a retrieval system, or transmitted, in any form or by any means (electronic, mechanical, photocopying, recording, or otherwise), without the prior written permission of the publisher.

09 10 11 12 13/PMPH/9 8 7 6 5 4 3 2 1

ISBN-13: 978-1-60795-032-5
ISBN-10: 1-60795-032-4
Printed in China by People's Medical Publishing House of China
Copyeditor/Typesetter: Spearhead; Cover Designer: Mary McKeon

**Library of Congress Cataloging-in-Publication Data on File**

Sales and Distribution

*Canada*
McGraw-Hill Ryerson Education
Customer Care
300 Water St
Whitby, Ontario L1N 9B6
Canada
Tel: 1-800-565-5758
Fax: 1-800-463-5885
www.mcgrawhill.ca

*Foreign Rights*
John Scott & Company
International Publisher's Agency
P.O. Box 878
Kimberton, PA 19442
USA
Tel: 610-827-1640
Fax: 610-827-1671

*Japan*
United Publishers Services Limited
1-32-5 Higashi-Shinagawa
Shinagawa-ku, Tokyo 140-0002
Japan
Tel: 03-5479-7251
Fax: 03-5479-7307
Email: kakimoto@ups.co.jp

*United Kingdom, Europe, Middle East, Africa*
McGraw Hill Education
Shoppenhangers Road
Maidenhead
Berkshire, SL6 2QL
England
Tel: 44-0-1628-502500
Fax: 44-0-1628-635895
www.mcgraw-hill.co.uk

*Singapore, Thailand, Philippines, Indonesia,*
*Vietnam, Pacific Rim, Korea*
McGraw-Hill Education
60 Tuas Basin Link
Singapore 638775
Tel: 65-6863-1580
Fax: 65-6862-3354
www.mcgraw-hill.com.sg

*Australia, New Zealand*
Elsevier Australia
Locked Bag 7500
Chatswood DC NSW 2067
Australia
Tel: 161 (2) 9422-8500
Fax: 161 (2) 9422-8562
www.elsevier.com.au

*Brazil*
Tecmedd Importadora e Distribuidora
de Livros Ltda.
Avenida Maurilio Biagi 2850
City Ribeirao, Rebeirao, Preto SP
Brazil
CEP: 14021-000
Tel: 0800-992236
Fax: 16-3993-9000
Email: tecmedd@tecmedd.com.br

*India, Bangladesh, Pakistan, Sri Lanka, Malaysia*
CBS Publishers
4819/X1 Prahlad Street 24
Ansari Road, Darya Ganj, New Delhi-110002
India
Tel: 91-11-23266861/67
Fax: 91-11-23266818
Email:cbspubs@vsnl.com

*People's Republic of China*
PMPH
Bldg 3, 3rd District
Fangqunyuan, Fangzhuang
Beijing 100078
P.R. China
Tel: 8610-67653342
Fax: 8610-67691034
www.pmph.com

Notice: The authors and publisher have made every effort to ensure that the patient care recommended herein, including choice of drugs and drug dosages, is in accord with the accepted standard and practice at the time of publication. However, since research and regulation constantly change clinical standards, the reader is urged to check the product information sheet included in the package of each drug, which includes recommended doses, warnings, and contraindications. This is particularly important with new or infrequently used drugs. Any treatment regimen, particularly one involving medication, involves inherent risk that must be weighed on a case-by-case basis against the benefits anticipated. The reader is cautioned that the purpose of this book is to inform and enlighten; the information contained herein is not intended as, and should not be employed as, a substitute for individual diagnosis and treatment.

www.pmph-usa.com/sperber

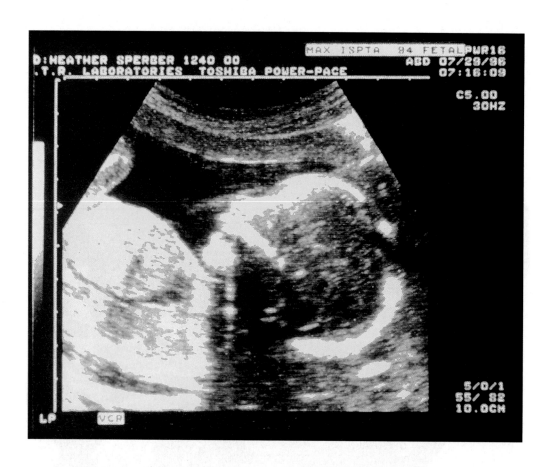

Ultrasonography has converted embryology and fetology from esoteric basic sciences into clinically relevant disciplines by making the fetus a potential patient. An ultrasonograph of the head and thorax of a normal 18-week-old fetus in utero.

# FOREWORD

Thirty-five years have elapsed since Emeritus Professor Geoffrey H. Sperber gave the world of dentistry and odontology the first edition of what was to prove his best-selling handbook on the development of morphology of the face and calvaria, jaws and teeth. The work has run through a number of editions, including Japanese, Indonesian and German translations. Those who are interested as well in phylogenetic development have noted how Sperber's book itself has evolved. The first four editions bore the title *Craniofacial Embryology*. An essentially new book appeared in 2001 under the title *Craniofacial Development*. This was more than a change of title, for it connoted a transformed philosophy and technology. Molecular biology was revolutionizing the study of ontogenetic development and of maldevelopment. Increasingly refined techniques made possible the rise of clinical diagnosis and eventually intervention in the fetus and even the embryo. Clinical embryology emerged in the latter part of the 20th century and is flourishing with the millennial debut.

Now, joined by his geneticist son, Steven Sperber, a new avatar emerges. The new Sperber book, *Craniofacial Embryogenetics and Development*, signifies the revolutionary and predominant rôle that genetics has come to play in the basic and clinical sciences. More than half a century has elapsed since James Watson and Francis Crick revealed the double-helix architecture of DNA and cracked the human genetic code. The Herculean task of mapping, not just a few obtrusive genes, but the total human genome, has been accomplished. It proved, in the end, to be not so everlastingly laborious and Sisyphean as had been surmised previously. Although it turned out to be numerically and quantitatively simpler, the heuristic value of the resulting fresco is already proving itself in myriad different applications. The greater simplicity of the emergent pattern surprised everyone but then, didn't Theodosius Dobzhansky tell us more than 50 years ago, "An extreme simplicity is characteristic of most of the really important conclusions in science"?

The moment when Geoffrey and Steven Sperber have been led to this exposition of the genes that influence facial development and maldevelopment is an auspicious one in the history of developmental and evolutionary studies. When this book appears in 2010, the world will have celebrated the bicentenary of Charles Darwin's birth (1809) and the 150th anniversary of the appearance of his greatest work, *The Origin of Species by Means of Natural Selection* (1859). For those who find Lamarckism to be philosophically diverting, 2009 saw the 200th anniversary of the appearance of Jean-Baptiste Lamarck's *Philosophie Zoologique* (1809). Alas, poor Lamarck and poor Darwin knew no genetics! While on the subject of anniversaries—which I love—in 2008, as I write these paragraphs, I cannot help mentioning that on July 1, 1858, there occurred a memorable meeting in the rooms of the Linnaean Society in London. Charles Darwin and Alfred Russel Wallace each presented his own paper on the theory of Natural Selection: So we have just enjoyed another 150th anniversary! The two men had independently arrived at Natural Selection as the key to the origin of species. At that time, Gregor Johann Mendel— whom some call "The Father of Genetics"—was carrying out his seminal experiments with garden peas in the Augustinian Monastery at Brno, Moravia, in what is today part of the Czech Republic. There is, however, no evidence that either Darwin or Wallace knew about Mendel's laws of inheritance, which he presented to the Society for the Study of Natural Science of Brno in 1865, and which were published in the Society's *Transactions* in 1866. They were then to all intents and purposes lost to the world of learning for 34 years. It was not until 1900 that Correns, De Vries and Tschermak independently obtained similar results to those of Mendel and, searching the literature, independently rediscovered the long-forgotten article of Mendel. By then much had been gleaned about the chromosomes. It was soon realized that their behavior during cell division, and especially in the "reduction divisions," or meiosis, that occurred during the formation of spermatozoa and ova, closely matched the behaviour of Mendel's essences that he had inferred from his crossing experiments on garden peas. Almost immediately, the way was paved for the new biological synthesis of the first third of the 20th century.

The new synthesis turned on a natural selective focus. It became known as the Neo-Darwinian Synthesis. One can but marvel at how far our ancestors—like Darwin and Wallace— had progressed in a pregenetical era!

Now we find ourselves in a genetic or even postgenetic era and this is the critical moment for Sperber and Sperber to apply the "new" genetical concepts to the development of the cranium and, especially, of the face, built around teeth and jaws. There are literally scores, if not a few hundred, genetic traits that express themselves in the myriad traits that go to make up the dentognathic system. In molding the dentition, for instance, genes apparently go to work directly on the mesenchymal dental pulp, adding a bit here, retarding a morsel there, while directing the development and patterning of the overlying enamel crowns. Mutant genes determine gradients of dental size and shape from mesial to distal; such niceties as reduced (peg-shaped) or even missing upper lateral incisors; shovel-shaped incisors; Carabelli traits; upper central diastema; agenesis of wisdom teeth and others; and shape of maxillary and mandibular dental arcades. For each of these curiosa, which are of greater or lesser clinical significance, there is assuredly a genetical underpinning. For some we cannot rule out epigenetic influences affecting, say, the enamel organs during ontogeny.

To these variegated musings the new book by the duo of Geoffrey and Steven Sperber now joined by Geoffrey Guttmann has given rise. I am confident that it will prove as great and sustainable a success as those books that have gone before under the authorship of Emeritus Professor Geoffrey Hilliard Sperber, anatomist, embryologist, anthropologist, and odontologist extraordinaire.

Phillip V. Tobias
FRS, FRCP, DSc,PhD, MBBCh

# ACKNOWLEDGMENTS

This book could not have been accomplished without the inspiration and unstinting assistance of numerous colleagues and helpers to whom I am deeply indebted. Many of the diagrams appearing in the predecessor *Craniofacial Embryology* drawn by Dr. Anthony M. MacIsaac have been reproduced, to which the artistic skills of Mrs. Jackie Wald have been added.

Special thanks go to Beth and Scott Lozanoff for their front cover portrayal of craniofacial embryogenetics. I am indebted to Dr. Geoffrey Machin for contributing Chapter 20.

My thanks go to the authors and publishers of diagrams and photographs, acknowledged in the captions, who so generously gave me permission to reproduce their copyrighted material in this work. The copyright permissions granted by W. B. Saunders, Springer-Verlag, Lippincott-Raven Publishers, Mosby, and Butterworth-Heinemann are gratefully acknowledged. Figures 1-5, 2-1, 2-2, 10-1, 10-4, and 10-7 published in Losee and Kirschner's *Comprehensive Cleft Care* 2009 by the McGraw-Hill Companies were reproduced by their kind permission.

My appreciation is extended to Professor Emeritus Phillip V. Tobias for providing the Foreword and to Dr. Geoffrey Guttmann for producing computer-created three-dimensional images of sectioned embryos.

My affectionate thanks to my wife, Robyn, and my daughters Heather and Jacqueline and their families for forgiving of my time spent in the preparation of this work.

G.H. Sperber

# PREFACE

The field of embryology has experienced a period of explosive growth since the previous edition of this book was published a decade ago. The insights of genetic expression in determining the unfolding of the embryonic layers have revolutionized our understanding of some of the mechanisms of embryogenesis. By linking genetics with embryology, the rationale for the title change of this book from *Craniofacial Development* to *Craniofacial Embryogenetics and Development* can be legitimized. Virtually all embryological development has an underlying genetic component, and genetics is a key basic science in uncovering the many mysteries of embryogenesis. The tools of molecular genetics have provided insights into developmental mechanisms that allow the ability to identify transient regions of genetic expression patterns. Unraveling the precise biochemical and mechanical interactions of discrete regions in the unfolding embryonic components remains a dauntingly complex challenge to understanding the conversion of the genome into the phenome. The addition of genetic information gleaned from other mammalian species might aid in dissecting human embryology into comprehensible components to understand normal and abnormal development. To this objective is this book dedicated.

G.H. Sperber
December 2008

# PREFACE TO PREVIOUS EDITION

The genesis of this book can be traced to the four editions of its predecessor *Craniofacial Embryology* and the successive *Craniofacial Development* that have enlightened students in the intricacies of development of the cephalic region for over a quarter of a century. The translation of these previous works into Japanese, German and Indonesian editions indicate widespread interest in utilizing this text in many parts of the world.

Replacement of that now outdated text is necessitated by the new technologies that have revealed deeper insights into the mechanisms of embryogenesis. The completed human genome project, and the explosion of molecular biological knowledge are expanding exponentially our understanding of the development and maldevelopment of the human organism. Moreover, the realm of the once esoteric study of embryology has been vaulted into clinical conscience now that *in vitro* fertilization, choronic villus sampling, amniocentesis and prenatal ultrasonography and even prenatal fetal surgery are becoming part of clinical practice (Frontispiece). All these aforementioned procedures require an intimate knowledge of the different stages of development, to which this book is dedicated.

The increasing sophistication of prenatal imaging techniques is revealing ever-earlier stages of fetal formation, and significantly, malformation, that may require intervention. To that end, increased emphasis has been placed on anomalous development, allowing for informed decision-making on possible reparative or "genetically engineered" therapy. The combining of "basic science" embryology and fetology with its consequences on clinical practice is one of the aims of this text in breaking barriers between "scientists" and "clinicians" in advancing our understanding the causes, prognosis and treatments of dysmorphology that is becoming an extensive component of modern medicine and its allied professions.

The advent of computer technology has enabled the portrayal of developmental phenomena as three-dimensional model images in sequential depictions of changes proceeding in the fourth dimension of time. This "morphing" technique is a powerful adjunct to gaining understanding of the complexities of rapidly changing tissues, organs and relationships occurring during embryogenesis. Some of these animated sequences have been added as an electronic adjunct to the text in the accompanying CD-ROM diskette.

It is hoped that the disparate disciplines of anatomy, embryology, syndromology, plastic and orofacial surgery, otolaryngology, orthodontics, pediatrics, dentistry, speech pathology and associated health care fields will find a melding of their diverse interests in the common objective of understanding and providing diagnosis, prognosis, prevention and cure of the disparities of development of the craniofacial complex. To this purpose is this book committed.

G.H. Sperber
October 2000

# CONTENTS

# SECTION I

## GENERAL EMBRYOLOGY

# 1 Mechanisms of Embryology

*"There must be a beginning of any great matter, but the continuing unto the end until it be thoroughly finished, yields the true glory."*

Sir Francis Drake (1587)

Unraveling the incredibly complex combination of molecular events that constitutes the creation of a human embryo is a form of ultradissection at the biochemical, cellular, organismal and systemic levels at different stages of development. The reconstitution of these unraveled discrete events provides a basis of understanding of the mechanisms of embryogenesis.

The new reproductive technological revolution instigated by laboratory in vitro fertilization (versus the old-fashioned in vivo method of fertilization) has sparked enormous advances in understanding the cascade of reactions initiated by the conjunction of viable spermotozoa with an ovum, whether it be abnormally in a Petri dish or normally in the oviduct. Insights into developmental phenomena are dependent upon a knowledge of genetics, of gene expression, of receptor mechanisms, of signal transduction and the differentiation of the totipotential stem cells into the different cell types that form tissues, organs and systems. Understanding these phenomena is changing embryology from a descriptive science into a predictive science with the potential for control of its mechanisms.

## GENETICS

The concept of a gene as director of development in conjunction with environmental influences needs to be understood in different contexts. The classic concept of a gene as a *unit* refers to a particular combination of DNA nucleotide base pairs, whereas consideration of a gene as a *unit of mutation* varies biochemically from a single base pair to hundreds of nucleotide base pairs. The embryologically significant gene as a *unit of function* is a sequence of hundreds or thousands of nucleotides that specify the sequence of amino acids that make up the primary structure of a polypeptide chain. These polypeptide chains constitute the proteins that provide the cells that form the tissues that create the organs of a developing embryo. Functionally, genes are conceived as structural, operator or regulatory genes. Analysis of an indivdual's genetic endowment is obtained by a metaphase spread of the chromosomes (Fig. 1–1) that is organized into a karyotype for analysis (Fig. 1–2). The pathway of interpretation of the genetic code into the phenotype is through transcription, translation and morphogenesis (Fig. 1–3).

The term *genome* refers to the array of genes (as above defined) in a complete haploid set of chromosomes that is expressed as the functional *genotype* in development that in combination with environmental influences

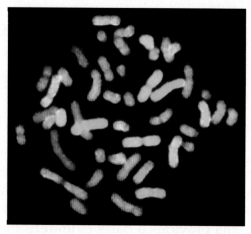

Figure 1–1 Spectral karyotyping (SKY) analysis of a normal human metaphase spread. Left: Inverted DAPI (4′-6-diamidino-2-phenylindole) image of the metaphase spread. The banding pattern is similar to G-banding pattern. Right: SKY analysis of the same metaphase spread. Red Green Blue display. (Courtesy of Dr. Bassem R. Haddad, Lombardi Comprehensive Cancer Center, Georgetown University, Washington, DC.)

results in formation of the *phenotype*, the physical and behavioral traits of an organism. The human genome, having been mapped (the Human Genome Project), is believed to contain $3 \times 10^9$ (3.1 billion) nucleotide base pairs that constitute approximately 20,000 genes*. The identification and mapping of these genes with specific positions (loci) on the chromosomes and their nomenclature is a task of increasing difficulty that is being standardized by the establishment of computerized Websites that are being continually updated as new data become available.

Regulation of the genetic program underlying cell differentiation and morphogenesis is due to differential gene activity. Only about 6% of human genes are made from a single linear piece of DNA. Most genes are made from coding regions, known as exons, found at different locations along a DNA strand. These data-encoded fragments are, by transcription, joined together and processed into a functional messenger RNA (mRNA) that forms a template for translation to generate proteins. Many genes encode *isoforms*, alternate forms of the gene transcripts arising from variations of the exons used for the coding sequence. This process emerges through "alternative splicing" and produces mRNA molecules and proteins of varying functions despite being formed from the same gene. Turning genes on and off at critical times determines cell fates, mitotic and apoptopic (cell death) activity, migratory patterns and metabolic states. There is a high degree of order in the genetic program, which is bolstered by redundancy and overlapping of expression patterns to guide morphogenesis. Intervention of developmental programs is the basis of experimental embryology and offers the potential for genetic engineering of deleterious or advantageous mutations. The rapid rate of cell division in the fetus may

*The size of the genome is independent of its genetic information. A single-cell ameba has a genome of >200,000 megabases; the human genome is about 3,000 megabases.

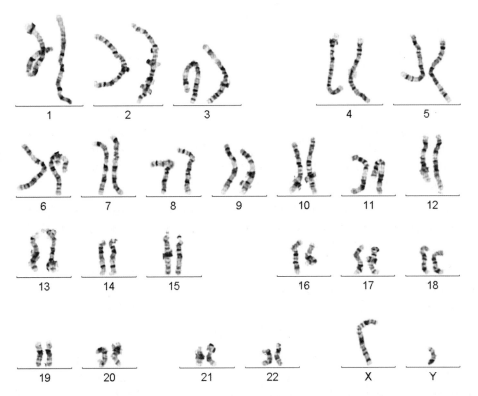

**Figure 1–2** A Normal male karyotype, showing the arranged chromosomes for analysis. (Courtesy of Quest Diagnostics, Chantilly, VA.)

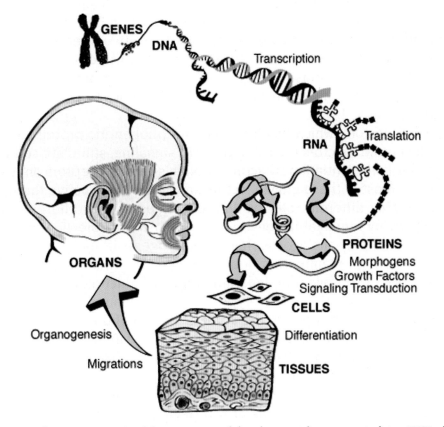

**Figure 1–3** Schematic synopsis of the sequence of development from genes to fetus (DNA, deoxyribonucleic acid; RNA, ribonucleic acid).

allow in utero vectored gene therapy for previously diagnosed mutations to correct genetic defects. *Epigenesis* describes the phenomena occurring after genetic determination, which provides an additional layer of regulatory complexity initially established by one's hereditary lineage and influenced by one's environment.

Genes control the synthesis of proteins of which some 1,000,000 varieties have been identified to create 200 or so cell types that proliferate into approximately $10^{14}$ cells forming 220 named structures in an average human adult. The longevity and proliferation of differentiated cells is also genetically determined in three broad categories:

1. Continuous mitotics (with short lifespans); for example, epithelia
2. Intermittent mitotics; for example, liver cells (hepatocytes)
3. Postmitotics (with long lifespans); for example, neurons

The advent of stem cell technology has added the possibility of regeneration of any of these cell types.

## SIGNAL TRANSDUCTION

Intercellular communication plays a major role in controlling development. Transcription factors regulate the identity and patterning of embryonic structures and the development of individual organs. Organizing centers are created that serve the source of signals that guide the patterning of organs and ultimately of the whole embryo. A signaling center or node (e.g., Hensen's node) is a group of cells that regulates the behavior of surrounding cells by producing positive and negative intercellular signaling molecules. Genes encode extracellular matrix proteins, cell adhesion molecules and cytoplasmic signaling pathway components. An ever-increasing number of signaling factors influencing development are being identified (Table 1–1).

Growth factors, which include bone morphogenetic proteins (BMPs), fibroblast growth factors (FGFs) and WNT* signaling, stimulate cell proliferation and differentiation by acting through specific receptors on responsive cells. Most of these growth factors are present and active throughout life, assuming different roles at different times and at different places, but displaying remarkable conservation of functional mechanisms. Thus, growth factors play analagous roles in embryogenesis, in the immune system and during inflammation and wound repair. This has given rise to the concept of "ontogenetic inflammation," by which normal embryonic development may act as a prototypic model for inflammation and healing that regulates homeostasis in the adult. Diffusion of these molecules and differential concentration gradients create fields of influence, determining differentiation patterns that form fate maps. After a signaling center has fulfilled its task, it gradually disappears.

Patterning of development of regions, of organs and of systems is controlled by genes expressed as growth factor–signaling molecules. The

*WNT: Drosophila Wingless and mouse homolog INT-I

TABLE 1–1: Signaling and Growth Factors

| Factor | Abbreviation | Derivation | Action |
|---|---|---|---|
| Homeodomain Transmission factors | HOXA, HOXB, PAX | Genome | Rostrocaudal and dorsoventral patterning |
| Sonic hedgehog protein | SHH | Axial mesendoderm | Neural plate patterning |
| Brain-derived neural growth factor | BDNF | Neural tube | Stimulates proliferation of neurons |
| Insulin-like growth factors I and II | IGF-I IGF-II | Liver | Regulator of somatic growth and cellular proliferation, inhibitor of apoptosis |
| Nerve growth factor | NGF | Various organs | Promotes axon growth and neuron survival |
| Transforming growth factor-$\alpha$ | TGF$\alpha$ | Various organs | Promotes differentiation of certain cells |
| Transforming growth factor-$\beta$ (Activin A, Activin B) | TGF$\beta$ | Various organs | Notochord and mesoderm induction. Potentiates or inhibits response to other growth factors. |
| Fibroblastic growth factor | FGF | Various organs | Notochord and mesoderm induction. Stimulates proliferation of fibroblasts, endothelium, myoblasts |
| Transcriptional factor families | TFs | DNA binding factors tightly spatiotemporally regulated | Stimulates transcription of downstream gene in cellular proliferation, differentiation, migration and apoptosis |
| Platelet-derived growth factor | PDGF | Various tissues | Stimulates proliferation of connective tissue cells and neuroglia |
| Epidermal growth factor | EGF | Various organs; salivary glands | Stimulates proliferation and differentiation of many cell types |
| Interleukin-2 | IL-2 | White blood cells | Stimulates proliferation of T lymphocytes |
| Hemopoietic cell growth factors Erythropoietin | IL-3, GM-CSF M-CSF, G-CSF | Hemopoeitic cells | Stimulates erythropoiesis |
| Vascular endothelial growth factor | VEGF | Various tissues | Stimulates angiogenesis |

development of a *primitive streak*† demarcates the initial distinction of embryonic tissues. A gene, Lim-1, is essential for organization of the primitive streak and development of the entire head. After the initial differentiation of the primary germ layers by reciprocal interactions between cells and tissues, segmentation is a feature of early embryogenesis. Such segmentation is expressed in the ectodermal neural tube into four regions: forebrain,

†A rapidly proliferating elongating mass of cells in the embryonic germ disk.

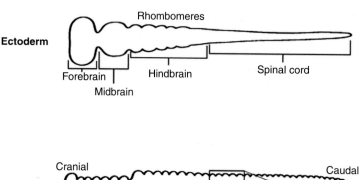

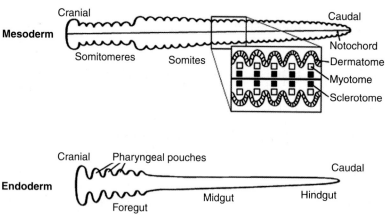

**Figure 1–4** Schematic depiction of segmentation in early embryogenesis in the primary germ layers.

midbrain, hindbrain and spinal cord, with the hindbrain being further segmented into seven or eight *rhombomeres*. Paraxial mesoderm is segmented cranially into seven swellings termed *somitomeres* and caudally into 38 to 42 *somites*. The six pharyngeal arches are a third visibly segmented set of tissues. (Fig. 1–4). Under the control of regulatory homeobox genes (Hoxa-1, Hoxa-2, Hoxb-1, Hoxb-3, Hoxb-4, Sonic hedgehog [SHH], Krox-20, Patched [Ptc], Paired Box, [Pax.9]) the segmented tissues are integrated into morphologically identifiable structures. (Fig. 1–4).

The mapping of genetic loci and the identification of mutant genes related to congenital defects and clinical syndromes are revealing the mechanisms of morphogenesis. The etiology and embryopathogenesis of a number of diverse anomalies of development are being traced to a common molecular basis. The recognition of growth and signaling factors (intercellular mediators) and their receptors (specific binding agents) and the expression domains of genes are providing insights into the mechanisms of normal and anomalous development (Fig. 1–5).

The developmental ontogeny of the craniofacial odontostomatognathic complex is dependent primarily on the following three elements:

1. Genetic factors: inherited genotype, expression of genetic mechanisms
2. Environmental factors: Nutrition and biochemical interactions; physical phenomena—temperature, pressures, hydration, etc.
3. Functional factors: Extrinsic and intrinsic forces of muscle actions, space-occupying cavities and organs, growth expansion, atrophic attenuation.

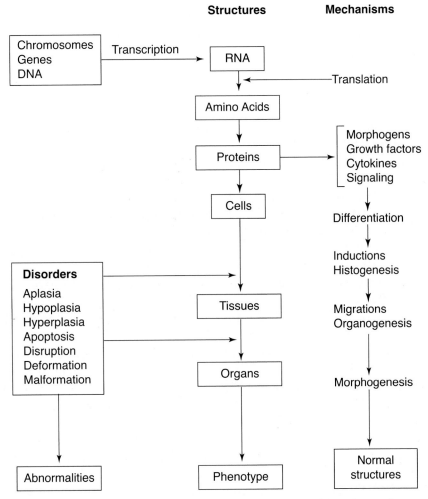

**Figure 1–5** Flow chart of developmental phenomena. (DNA, deoxyribonucleic acid; RNA, ribonucleic acid.)

The central theme of patient care, around which rotate therapies and researchers, provides interactive feedback for insights into developmental phenomena and their aberrations (Fig. 1–6).

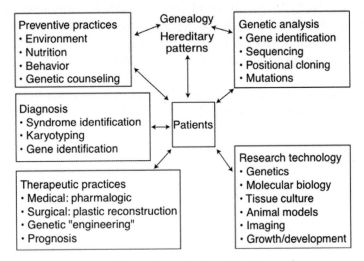

Figure 1–6 Schema for patient-centered procedures.

# SELECTED BIBLIOGRAPHY

Agarwal P, Verzi MP, Black BL. Genetic and functional interaction between transcription factors MEF2C and Dlx5/6 is required for craniofacial development. Dev Biol 2008; 319: 595.

Bruford EA. Human gene nomenclature. Encyclopedia of life sciences. New York, John Wiley & Sons, 2008.

Carlson BM. Human embryology and developmental biology. 4th ed. Philadelphia, Mosby Elsevier, 2009.

Chai Y, Maxson RE. Recent advances in craniofacial morphogenesis. Dev Dyn 2006; 235:2353–2375

Francis-West PH, Robson L, Evans DJR. Craniofacial development: the tissues and molecular interactions that control development of the head. Adv Anat Embryo Cell Biol 2003; 169:III–VI,1–138.

Handrigan GR, Buchtova M, Richman JM. Gene discovery in craniofacial development and disease—cashing in your chips. Clin Genet 2007; 71:109–119.

Hall BK. Germ layers and the germ layer theory revisited. Evolutionary Biol 1998; 30:121–186.

Hart TC, Marazita ML, Wright JT. The impact of molecular genetics on oral health paradigms. Crit Rev Oral Biol Med 2000; 11:26–56.

Mukhopadhyay P, Greene RM, Pisano MM. Expression profiling of transforming growth factor beta superfamily genes in developing orofacial tissue. Birth Defects Res (Part A) 2006; 528–543.

Hu D, Marcucio RS. Unique organization of the frontonasal ectodermal zone in birds and mammals. Dev Biol 2009; 325:200–210

Hu D, Marcucio RS. A SHH-responsive signaling center in the forebrain regulates craniofacial morphogenesis via the facial ectoderm. Development 2009; 136:107–116.

Hunt P, Clarke JDW, Buxton P, et al. Segmentation crest prespecification and the control of facial form. Eur J Oral Sci 1998; 106(Suppl):12–18.

Jeong J, Li X, McEvilly RJ, et al. Dlx genes pattern mammalian jaw primordium by regulating both lower jaw–specific and upper jaw–specific genetic programs. Development 2008; 135:2905–2916.

Kalyani AJ, Rao MS. Cell lineage in the developing neural tube. Biochem Cell Biol 1998; 76:1051–1068.

Kerrigan JJ, McGill JT, Davies JA, et al. The role of cell adhesion molecules in craniofacial development. J R Coll Surg Edin 1998; 43:223–229.

Korshunova Y, Tidwell R, Veile R, et al. Gene expression profiles of human craniofacial development.COGENE: http://hg.wustl.edu.COGENE. Accessed March 3, 2009.

Le Douarin NM, Brito JM, Creuzet S. Role of the neural crest in face and brain development. Brain Res Rev 2007; 55:237–247.

Mao J, Nah H-D. Craniofacial growth and development. Ames, Iowa, Wiley-Blackwell, 2010.

Moore KL, Persaud TVN. The developing human. 8th ed. Philadelphia, Saunders Elsevier, 2008.

Mukhopadhyay P, Greene RM, Zacharias W, et al. Developmental gene expression in profiling of mammalian fetal orofacial tissue. Birth Defects Res A Clin Mol Teratol 2004; 70:912–926.

Opitz JM. Blastogenesis and the "primary field" in human development. Birth Defects: Orig Art Ser 1993; 29:3–37.

Pirinen S. Genetic craniofacial aberrations. Acta Odont Scand 1998;56:356–359.

Richman JM, Rowe A, Brickell PM. Signals involved in patterning and morphogenesis of the embryonic face. Prog Clin Biol Res 1991; 373:117–131.

Rossel M, Capecchi MR. Mice mutant for both hoxa1 and hoxb1 show extensive remodeling of the hindbrain and defects in craniofacial development. Development 1999; 126:5027–5040.

Sela-Donenfeld D, Kalcheim C. Regulation of the onset of neural crest migration by coordinated activity of BMP4 and Noggin in the dorsal neural tube. Develop 1999; 126:4749–4762.

Schoenwolf GC, Bleyl SB, Brauer PR, Francis-West PH. Larsen's human embryology. 4th ed. New York,.Churchill Livingstone, 2009.

Shuler CF. Programmed cell death and cell transformation in craniofacial development. Crit Rev Oral Biol Med 1995; 6:202–217.

Smith CM, Finger JH, Hayamizu TF, et al. The Gene Expression Database (GXD): A resource for developmental biologists. Dev Biol 2008; 319:564.

Sperber GH. First year of life: prenatal craniofacial development. Cleft Palate-Craniofac J 1992; 29:109–111.

Sperber GH. Current concepts in embryonic craniofacial development. Critical Rev Oral Biol Med 1992; 4:67–72.

Sperber GH, Machin GA. The enigma of cephalogenesis. Cleft Palate-Craniofac J 1994; 31:91–96.

Stewart CL, Cullinan FB. Preimplantation development of the mammalian embryo and its regulation by growth factors. Dev Genet; 1997; 21:91–101.

Thesleff I. The genetic basis of normal and abnormal craniofacial development. Acta Odont Scand 1998; 56:321–325.

Wood R, ed. Genetic nomenclature guide with information on websites. Trends Genet Suppl 1998. Elsevier pages 1–49.

Wilkie AOM, Morriss-Kay GM.Genetics of craniofacial development and malformation. Nature Rev Genet 2001; 2(6);458–468.

Wozney JM. The bone morphogenetic protein family: multifunctional cellular regulators in the embryo and adult. Eur J Sci 1998; 106:160–166.

## Websites

www.gene.ucl.ac.uk/nomenclature
www.ncbi.nih.gov/genome/guide
http://www.hgmd.cf.ac.uk/
http://www.genenames.org
www.genepaint.org
www.cmbi.ru.nl/GeneSeeker/
http://hgwustl.edu/COGENE/
http://genome.ucsc.edu/
http://www.ensembl.org/Homo_sapiens/Info/Index
http://www.ncbi.nlm.nih.gov/genome/guide/human/
www.med.unc.edu/embryo_images/
http://virtualhumanembryo.lsuhsc.edu/

# 2 Early Embryonic Development

*Over the structure of the cell rises the structure of plants and animals, which exhibit the yet more complicated, elaborate combinations of millions and billions of cells coordinated and differentiated in the most extremely different way.*

Oscar Hertwig

The mating of male and female gametes in the maternal uterine tube initiates the development of a zygote—the first identification of an individual. The union of the haploid number of chromosomes (23) of each gamete confers the hereditary material of each parent upon the newly established diploid number of chromosomes (46) of the *zygote*. All the inherited characteristics of an individual and its sex are thereby established at the time of union of the gametes. The single totipotential cell of approximately 140 μm diameter resulting from the union very soon commences mitotic division to produce a rapidly increasing number of smaller cells, so that the 16-cell stage, known as the *morula*, is not much larger than the initial zygote. These cells of the early zygote reveal no significant outward differences of form. However, the chromosomes of these cells must necessarily contain the potential for differentiation of subsequent cell generations into the variety of cell forms that later constitute the different tissues of the body. The genetic material contained in the reconstituted diploid number of 46 chromosomes of the initial zygote cell, by replication, is identical to that in its progeny. The activity of this replicated genetic material varies as the subsequent cell generations depart from the archetypical "primitive" cell. Parts of the genetic material are active at certain stages of development, whereas other parts that might remain quiescent at those particular times become active at others. Proliferation of the cells of the zygote allows expression of their potential for differentiation into the great variety of cell types that constitute the different tissues of the body. The differentiation of these early pluripotential cells into specialized forms is dependent upon genetic, cytoplasmic and environmental factors that act at critical times during their proliferation and growth.

## DIFFERENTIATION

The transformation of the ovum into a fully fledged organism by which process there is orderly enlargement and diversification of the proliferating cells of the morula is the result of selective activation and repression of the diploid set of genes carried in each cell. Which one of a pair of gene alleles contained in the diploid set of autosomal chromosomes is expressed depends upon their similarity (homozygosity) or dissimilarity (heterozygosity). In the latter case, the degree of dominance or recessiveness of each allele of the pair determines the phenotypic expression of the gene. The expression of the traits governed by genes on the pair of nonautosomal or sex chromosomes is somewhat different. In females there is inactivation of one of the two X chromosomes (termed a Barr body) and failure of expression of its genes (the Lyon hypothesis). In males the presence of the Y chromosome with its sex-determining region (SRY) and only a single X chromosome accounts for the sex linkage of certain inherited traits.

A programmed sequence of development, known as *epigenesis,* is dependent upon determination that restricts multipotentiality and causes *differentiation* of proliferating cells. These developmental events result from continuous interactions between cells and their microenvironments. As a consequence of differentiation, new varieties of cell types and tissues develop that interact with one another by *induction,* producing an increasingly complex organism. Induction alters the developmental course of responsive tissues, whose capacity to react is known as *competence,* to produce different tissues from which organs and systems arise. Inductive interactions may take place in several ways in different tissues. Interactions may occur by direct cellular contact, or may be mediated by diffusible agents, or even by inductors enclosed in vesicles. The mechanisms involved in these processes include gene activation and inactivation, protein translation mechanics, varying cell membrane selectivities, intercellular adhesions and repulsions and cell migration that produce precise cell positioning. Cell position and adhesion are key factors in early morphogenesis, as microenvironments activate or inhibit mechanisms leading to cellular diversification. All these events are critically timed and are under hormonal, metabolic and nutritional influences. The biochemical foundations of these complex functions, and the nature and manner of operation of their controlling factors, which are being widely explored, are among the central challenges of contemporary biology. The identification of *morphogens* determining differentiation and *teratogens* disturbing normal morphogenesis is the current focus of developmental biologists.

Units of cells and tissues form *morphogenetic fields,* which follow genetic and epigenetic phases of morphogenesis. Fields are susceptible to alteration by interplay of genetic and environmental factors. The peak period of morphogenesis of a developmental field is a critical period of sensitivity to environmental and teratogenic disturbances. A compendium of manifold biochemical reactions leads to *cytodifferentiation* and *histodifferentiation,* resulting in the formation of epithelial and mesenchymal tissues that acquire specialized structure and function (Fig. 2–1). Epithelial-mesenchymal interactions that provide for reciprocal cell differentiation are essential to organogenesis, that is, the production of organs and systems (Fig. 2–2).

The entire group of the above processes is marvelously integrated to form the external and internal configuration of the embryo, constituting *morphogenesis,* the process of development of form and size, that determines the morphology of organs and systems and the entire body. Not only is mitosis and cell growth essential for embryonic development, but paradoxically, even cell death—genetically and hormonally controlled—forms a significant part of normal embryogenesis. By means of programmed cell

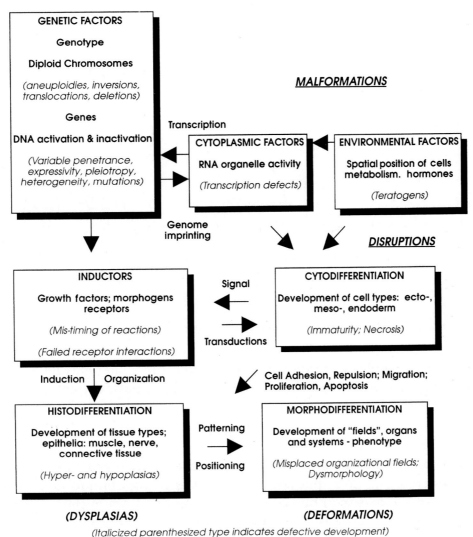

**Figure 2–1** Schema of embryogenesis (and possible sources of anomalies).

death, tissues and organs useful only during embryonic life are eliminated, along with phylogenetic vestiges developed during ontogeny.

The expressed character of differentiated cells (the phenotype) will depend first on their genetic constitution (the genotype) and, secondly, on the type and degree of gene expression and repression and of environmental influences during differentiation. Defective genes (mutations) or abnormal chromosome numbers (aneuploidy or polyploidy) will pattern aberrant development. Adverse environmental factors, both prenatal and postnatal, can cause the genotypic pattern to deviate from normal development. Neither heredity (the genotype) nor the environment ever works exclusively in patterning development, but always in combination to produce the phenotype.

Disturbances of the inductive patterns of embryonic tissues will result in congenital defects of development. *Teratology* constitutes the study of such abnormal development. Malformations of the face and jaws are

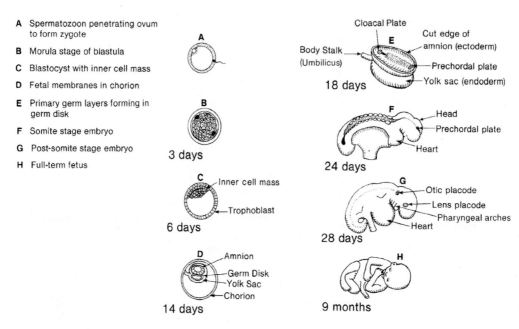

**A** Spermatozoon penetrating ovum to form zygote

**B** Morula stage of blastula

**C** Blastocyst with inner cell mass

**D** Fetal membranes in chorion

**E** Primary germ layers forming in germ disk

**F** Somite stage embryo

**G** Post-somite stage embryo

**H** Full-term fetus

Figure 2–2  Schematized synopsis of salient features of general embryology.

frequently part of congenital abnormality syndromes that may be amenable to surgical, orthopedic, orthodontic and therapeutic correction.

## GROWTH

Growth is a fundamental attribute of developing organisms. The dramatic increase in size that characterizes the living embryo is a consequence of (1) increased number of cells resulting from mitotic divisions (hyperplasia); (2) increased size of individual cells (hypertrophy); and (3) increased amount of noncellular material (accretion). Hyperplasia tends to predominate in the early embryo, whereas hypertrophy largely prevails later. Once differentiation of a tissue has been established, further development is predominantly that of growth. The rate of growth of tissue is inherently determined, but is, of course, also dependent upon environmental conditions. The health, diet, race and sex of an individual influence the rate and extent of growth.

Growth may be *interstitial*, where increase in bulk occurs within a tissue or organ, or appositional, where surface deposition of tissue enlarges its size. Interstitial growth characterizes soft tissues, whereas hard tissues (bone, dental tissues) necessarily increase by apposition.

Growth is not merely an increase in size. If it were, the embryo would expand like a balloon, and the adult would simply be an enlarged fetus. The resulting unproportioned growth would produce a grossly distorted adult with a head as large as the rest of the body. Not all tissues, organs and parts develop at the same rate, *differential* growth accounting for a varying proportioned increase in size. The head is precocious in its development, constituting half the body size in the fetus but later undergoing a relative decrease in relation to total body size. Each organ system grows at its own predetermined rate, accounting for proportional discrepancies in size at different periods of life. Some organ systems enlarge precociously and, subsequently, remain nearly stationary in size, whereas others continue to grow until adolescence. The lymphoid system continues to grow until adolescence. The lymphoid system (tonsils, thymus, etc.), after extremely rapid growth in early childhood, even regresses in size before adulthood (Fig. 2–3).

Increments in growth are constantly changing with chronological age, being most rapid in the fetus and infant and again at puberty. Despite the varying growth rates of different organ systems, there is an overall harmony of proportions. As an example, teeth are initiated and grow at the precise time that the jaws have reached a size ready to accommodate them.

The apparently scattered order of eruption of the teeth is another manifestation of the phenomenon of differing rates of growth. The various categories of teeth erupt at different times. The growth and development sequence is genetically determined and operates through the mechanism of inductors, metabolic modulators, neurotrophic and hormonal substances and interacting systems of contact stimulation and inhibition of contiguous tissues. Should these differential, but integrated, rates of development fail to maintain their normal determined pace, aberrations of overall development will manifest themselves as malformations that may require clinical interception for correction. An example of such a human disorder is the very rare Proteus syndrome, causing overgrowth of tissues leading to asymmetry of most prominently the face, limbs, hands and feet through disproportionate growth rates.

Maturation is a counterpart of growth; it indicates not only the attainment of adult size and proportions but also the full adult constituents of

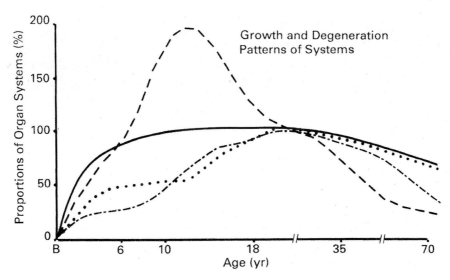

Figure 2–3 Graphic representation of the varying proportions of the various organ systems in postnatal life, where 100% represents the size of the organ systems in the mature young adult. _ _ _ _ _, Lymphoid system; _____, central nervous system; _._._._, dentition;  . . . . . . . ., general body. (Based on Scammon RE. The measurement of man. Minneapolis, University of Minnesota Press, 1930.)

tissues (e.g., mineralization) and the complete capability for performance of each organ's destined functions.

When the age of occurrence of maturational events is indicated (onset of ossification centers, fusion of sutures, eruption of teeth, etc.), it must be stressed that these manifestations of *biological* age of an individual need not coincide with *chronological* age, and, in fact, they often differ from one another. When biological age is well in advance of chronological age, the individual is developing rapidly; when the reverse occurs, the individual has a retarded rate of development.

Although most growth normally ceases at the end of adolescence, coinciding with the eruption of the third molar teeth (hence, the popular connotation of "wisdom" teeth), the facial bones, unlike the long bones, retain the potential for further appositional growth in adult life. Such postadolescent growth may occur as a result of hypersecretion of somatotrophic hormone from a pituitary gland tumor, as in acromegaly, which is characterized by enlargement of the bones of the face, hands and feet.

## PHASES OF DEVELOPMENT

Embryogenesis is divided into three distinct phases during the 280 days of gestation (10 28-day menstrual cycles*). The phases are the preimplantation period (the first 7 days), the embryonic period (the next 7 weeks), and the fetal period (the next 7 calendar months).

### Preimplantation Period

During the first 2 to 3 days postconception (pc), the zygote progresses from a single-cell to a 16-cell cluster, the morula, no larger than the original ovum (Fig. 2–4). The first 2 days of human embryo formation are controlled

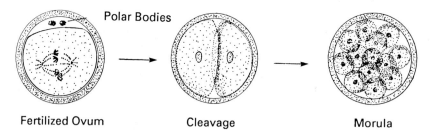

Polar Bodies

Fertilized Ovum          Cleavage          Morula

**Figure 2–4** Initial stages of embryogenesis, depicting cell division. Note that the morula, containing up to 16 cells, is no larger than the fertilized ovum.

---

*This duration of gestation is based upon the menstrual cycle of 28 days. Calculation from the last occurring menstruation is known as the "menstrual age" of the embryo. This menstrual age is most frequently used in obstetric practice, as it is based upon the last occurrence of an easily observed event. As ovulation and subsequent fertilization occur approximately 14 days after the last menstruation, the true age of the embryo ("fertilization age") is 2 weeks less than the menstrual age. Because they are more accurate, all ages referred to hereafter are fertilization ages unless otherwise specified. To distinguish from postnatal ages, prenatal ages are indicated as being postconception (abbreviated to pc).

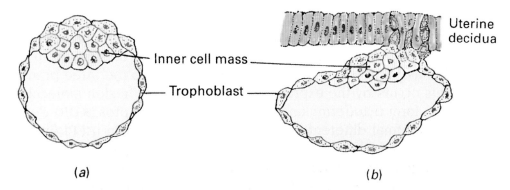

**Figure 2–5** The blastocyst (a), developed from the morula, implants (b) into the decidual layer of the uterine wall.

by factors expressed solely by the oocyte. Thereafter, the genetic factors of the biparental zygote are expressed in differentiation of the cells. The early totipotential blastomeres can develop into any tissue, but later differentiation creates an approximately 100-cell fluid-filled blastocyst. The outer sphere of cells forms the trophoblast, and the inner cell mass will form the embryo (Fig. 2–5). During this period, the conceptus passes along the uterine tube to enter the uterus, where it implants into the uterine endometrium on the seventh postconception day. The trophoblast converts into the chorion by sprouting villi. Chorionic implantation establishes the placenta, the organ of fetomaternal exchange of nutrition and waste disposal (Fig. 2–6).

## Embryonic Period

This phase, from the end of the first week until the eighth week, can be subdivided into three periods: presomite (8 to 21 days postconception),

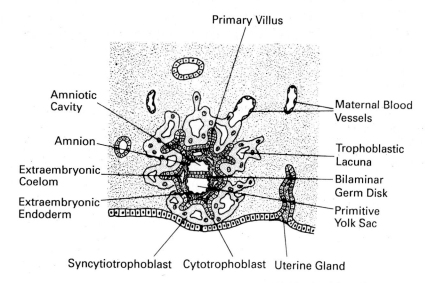

**Figure 2–6** Ten-day-old conceptus implanted in uterine wall. The fetal membranes, amnion, yolk sac and chorion (syncytiotrophoblast and cytotrophoblast) have formed. The forming placenta surrounds the conceptus.

somite (21 to 31 days postconception) and postsomite (32 to 56 days postconception). During the presomite period, the primary germ layers of the embryo and the embryonic adnexa (fetal membranes) are formed in the inner cell mass. Embryonic lineage differentiation into the three primary germ layers occurs by the expression of neural cell adhesion molecule 1 (NCAM1) to form ectoderm; kinase insert domain receptor (KDR) determines mesodermal differentiation, and alpha-fetoprotein (AFP) expression determines endoderm. In the somite period, characterized by the appearance of prominent dorsal metameric segments, the basic patterns of the main body systems and organs are established. The postsomite period is characterized by formation of the body's external features. The chronology of events during the embryonic period, divided into stages, is designated in Table 2–1.

### Fetal Period

The beginning of this long phase, from the 8th week until term, is identified by the first appearance of ossification centers and the earliest movements by the fetus. There is little new tissue differentiation or organogenesis, but there is rapid growth and expansion of the basic structures already formed.

## FETAL MEMBRANES*

Embryonic adnexa form membranes surrounding cavities in which the embryo (and subsequently the fetus) develops. These fluid-filled cavities, membranes and organs protect the fetus physically and serve its nutritional and waste-disposal requirements, and are cast off at birth. The main cavities and their membranes are the chorion and amnion, surrounding the fetus. Lesser cavities and their membranes are the yolk sac and a transient diverticulum, the allantois, that become incorporated into the umbilical cord, connected to the placenta, the chief organ of fetal sustenance (Figs. 2–6, 2–7, 2–10).

The chorion arises from the trophoblast as an all-encompassing membrane that, with the maternal endometrium, forms the placenta. The amnion and the yolk sac, fluid-filled sacs in the inner cell mass within the chorion, are separated by a bilaminar plate; this plate forms the embryonic disk, which later gives rise to the definitive embryo. The attachment of the inner cell mass to the chorion constitutes the connecting (body) stalk that contains the yolk sac and allantois. The stalk converts into the umbilical cord.

---

*The long-established term fetal membranes is a misnomer for the various extraembryonic structures collectively described under this title, as these structures are not necessarily "fetal" or "membranes." However, there is no suitable alternative nomenclature.

**TABLE 2–1:  Chronology of events during the embryonic period**

| Carnegie Stage | Postconception Age | Craniofacial Features |
|---|---|---|
| 6 | 14 days | Primitive streak appears; oropharyngeal membrane forms |
| 8 | 17 days | Neural plate forms |
| 9 | 20 days | Cranial neural folds elevate; otic placode appears |
| 10 | 21 days | Neural crest migration commences; fusion of neural folds; otic pit forms |
| 11 | 24 days | Frontonasal prominence swells; first arch forms; wide stomodeum; optic vesicles form; anterior neuropore closes; olfactory placodes appear |
| 12 | 26 days | Second arch forms; maxillary prominences appear; lens placodes commence; posterior neuropore closes; adenohypophyseal pouch appears |
| 13 | 28 days | Third arch forms; dental lamina appears; fourth arch forms; oropharyngeal membrane ruptures |
| 14 | 32 days | Otic and lens vesicles present; lateral nasal prominences appear |
| 15 | 33 days | Medial nasal prominences appear; nasal pits form, widely separated, face laterally |
| 16 | 37 days | Nasal pits face ventrally; upper lip forms on lateral aspect of stomodeum; lower lip fuses in midline; retinal pigment forms; nasolacrimal groove appears, demarcating nose; neurohypophyseal evagination |
| 17 | 41 days | Contact between medial nasal and maxillary prominences, separating nasal pit from stomodeum; upper lip continuity first established; vomeronasal organ appears |
| 18 | 44 days | Primary palate anlagen project posteriorly into stomodeum; distinct tip of nose develops; eyelid folds form; retinal pigment; nasal pits move medially; nasal alae and septum present; mylohyoid, geniohyoid and genioglossus muscles form |
| 19 | 47-48 days | Nasal fin disintegrates; (failure of disintegration predisposes to cleft lip); the rima oris of the mouth diminishes in width; mandibular ossification commences |
| 20 | 50-51 days | The lidless eyes migrate medially; nasal pits approaches each other; ear hillocks fuse |
| 22 | 54 days | The eyelids thicken and encroach upon the eyes; the auricle forms and projects; the nostrils are in definitive position |
| 23 | 56-57 days | Eyes are still wide apart but eyelid closure commences; nose tip elevates; face assumes a human fetal appearance; head elevates off the thorax; mouth opens; palatal shelves elevate; maxillary ossification commences |
| Fetus | 60 days | Palatal shelves fuse; deciduous tooth buds form; embryo now termed a fetus |

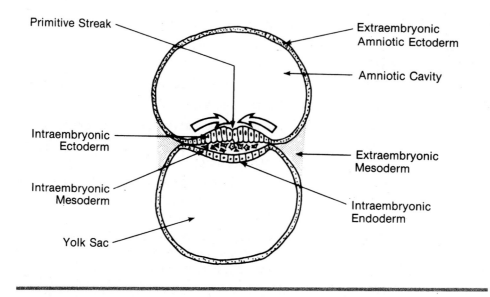

**Figure 2–7** Cross-sectional view of embryonic disk (center) and fetal membranes amnion (above) and yolk sac (below).

# EARLY EMBRYOGENESIS: EMBRYONIC PERIOD

## Presomite Period

*Development of the Endoderm and Ectoderm*

The primordial embryonic germ disk is composed of two primary germ layers: the ectoderm, which forms the floor of the amniotic cavity, and the endoderm, which forms the roof of the yolk sac (Figs. 2–7 and 2–8). There is early demarcation at the 14th day of the rostral pole of the initially oval disk by the appearance of a node. The Nodal, Hedgehog (HH), fibroblast

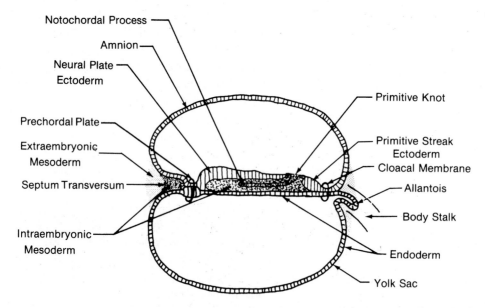

**Figure 2–8** Longitudinal section through a 14-day-old embryo depicting the fetal membranes (amnion, yolk sac, and allantois) and sites of ectoderm/endoderm contact (prechordal plate and cloacal membrane).

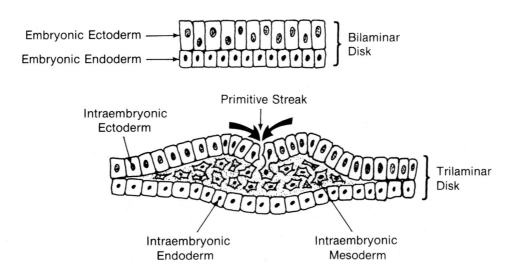

Embryonic Ectoderm ⟶ } Bilaminar Disk
Embryonic Endoderm ⟶

Primitive Streak

Intraembryonic Ectoderm

} Trilaminar Disk

Intraembryonic Endoderm      Intraembryonic Mesoderm

**Figure 2–9** Schematic cross-section of embryonic disk at bilaminar (above) and trilaminar (below) stages. Enlarged view of center of Figure 2–7.

growth factor (FGF), Wnt, bone morphogenetic protein (BMP) and Notch *signaling* pathways all converge to regulate the activity of Nodal. Morphogenesis of the node results in motile cilia, which create a leftward fluid flow that establishes left-right asymmetry in the subsequent embryo.

An endodermal thickening, the *prechordal plate\** appears in the future midcephalic region as a consequence of Sonic Hedgehog (SHH) *signaling* (Fig. 2–8). The prechordal plate prefaces the development of the orofacial region, giving rise later to the endodermal layer of the oropharyngeal membrane; the importance of this membrane is discussed later in relation to development of the mouth. The third primary germ layer, the mesoderm, makes its appearance at the beginning of the third week as a result of ectodermal cell proliferation and differentiation in the caudal region of the embryonic disk. The resultant bulge in the disk is grooved craniocaudally, by which characteristic it is termed the *primitive streak* (Fig. 2–9). From the primitive streak, the rapidly proliferating tissue known as mesenchyme forms the intraembryonic mesoderm which migrates in all directions between the ectoderm and endoderm except at the sites of the oropharyngeal membrane rostrally and the cloacal membrane caudally (Figs. 2–10 and 2–11). The appearance of the mesoderm converts the bilaminar disk into a trilaminar structure. The midline axis becomes defined by the formation of the notochord from the proliferation and differentiation of the cranial end of the primitive streak. The notochord serves as the axial skeleton of the embryo, and induces formation of the neural plate in the overlying ectoderm (neural ectoderm), and the lateral mesoderm induces epidermal development (cutaneous ectoderm) (Figs. 2–12 and 2–13).

---

\*The prechordal plate is believed to perform a head-organizing function. Defects in the prechordal plate may result in holoprosencephaly or agenesis of the corpus callosum. It gives rise to cephalic mesenchyme concerned with extrinsic eye muscle development. The prechordal plate also gives rise to the preoral gut (Seessel's pouch, qv).

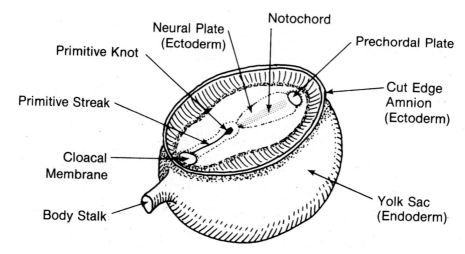

Figure 2–10  Dorsal surface view of the embryonic disk with the amnion cover removed. The prechordal plate demarcates the future mouth. The body stalk will form the umbilical cord connecting the embryo to the placenta.

The three primary germ layers serve as a basis for differentiating the tissues and organs that are largely derived from each of the layers. The cutaneous and neural systems develop from the ectodermal layer. The cutaneous structures include the skin and its appendages, the oral mucous membrane, and the enamel of teeth. The neural structures include the central and peripheral nervous systems. The mesoderm gives rise to the cardiovascular system (heart and blood vessels), the locomotor system (bones and muscles), connective tissues and dental periodontium. The endoderm develops into the lining epithelium of the respiratory system and of the alimentary canal between the pharynx and the anus, as well as the secretory cells of the liver and pancreas.

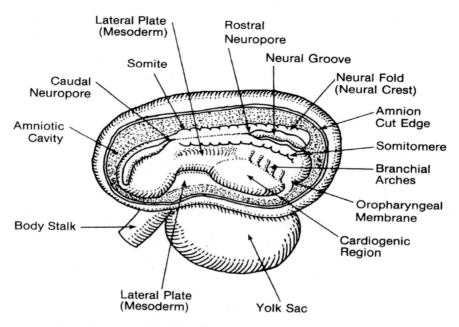

Figure 2–11  Lateral view of 23-day-old embryo depicting fusion of neural folds. The somites are forming.

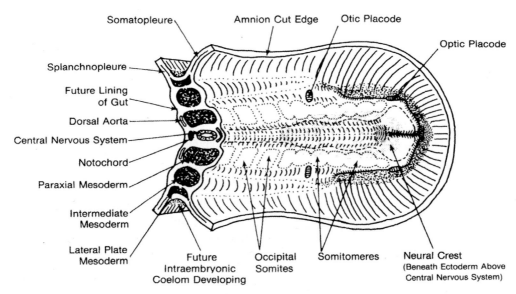

Figure 2–12 Composite "fate map" of dorsal surface and midsection of 23-day-old embryo.

Development of the ectoderm into its cutaneous and neural portions occurs at 20 days by infolding of the *neural plate* ectoderm along the midline forming the *neural folds*; this creates a *neural groove*. At 22 days, the neural folds fuse in the region of the third to fifth somites, the site of the future occipital region. This initial closure proceeds both cephalically and caudally to form the neural tube, which submerges beneath the superficial covering of the cutaneous ectoderm (see Figs. 2–11 and 2–12). Also, at this stage, the *lens* and *otic placodes*, which will form the eye lenses and inner ears, appear on the surface ectoderm. The anterior and posterior neuropores close at 25 and 27 days, respectively.

### Development of the Neural Crest

This ectomesenchymal tissue, termed the *neural crest* from its site of origin (see Fig. 2–13), arises from the crests of the neural fold where neuralizing and epidermalizing influences interact. Presumptive neural crest tissue broadly correlates with BMP expression domains during gastrulation of the embryonic disk. Neural crest cells form a separate tissue that, being akin to the three primary germ layers, is pluripotential. The neural crest is induced by a combination of secreted inductive signals from WNT activation and BMP inhibition. Later maintenance of the neural crest requires activation of both pathways. Neural crest ectomesenchyme possesses great migratory propensities, following natural cleavage planes, between mesoderm, ectoderm and endoderm, and tracking intramesodermally. These population shifts may be either passive translocations resulting from displacement of tissue or active cell migrations. Translocated neural crest cells, upon reaching their predetermined destinations, undergo cytodifferentiation into a wide variety of diverse cell types that are in part genetically predetermined and in part specified by local environmental influences (Table 2–2). Specification of cranial neural crest tissue into a

(*text continues on page* 29)

## Primary Germ Layers

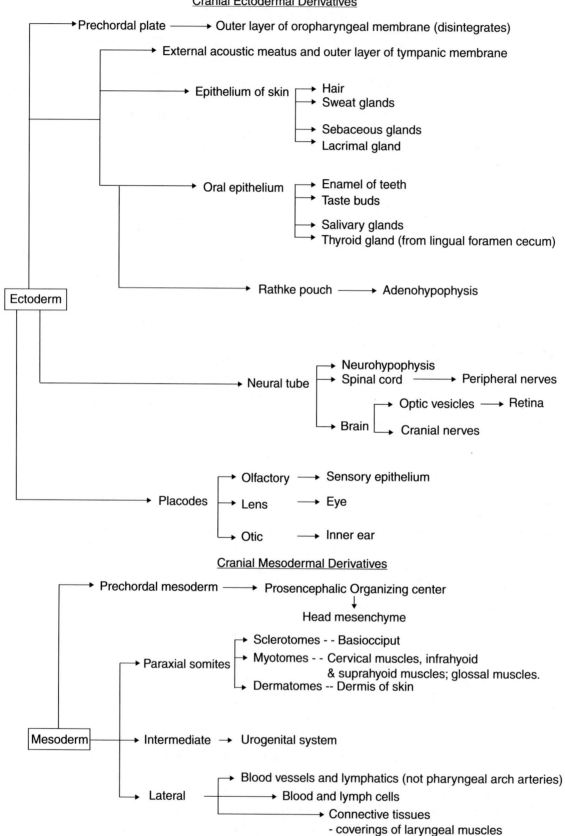

Cranial Ectodermal Derivatives

Prechordal plate ⟶ Outer layer of oropharyngeal membrane (disintegrates)

External acoustic meatus and outer layer of tympanic membrane

Epithelium of skin
- Hair
- Sweat glands
- Sebaceous glands
- Lacrimal gland

Oral epithelium
- Enamel of teeth
- Taste buds
- Salivary glands
- Thyroid gland (from lingual foramen cecum)

Rathke pouch ⟶ Adenohypophysis

Ectoderm

Neural tube
- Neurohypophysis
- Spinal cord ⟶ Peripheral nerves
- Brain
  - Optic vesicles ⟶ Retina
  - Cranial nerves

Placodes
- Olfactory ⟶ Sensory epithelium
- Lens ⟶ Eye
- Otic ⟶ Inner ear

Cranial Mesodermal Derivatives

Prechordal mesoderm ⟶ Prosencephalic Organizing center
↓
Head mesenchyme

Paraxial somites
- Sclerotomes - - Basiocciput
- Myotomes - - Cervical muscles, infrahyoid & suprahyoid muscles; glossal muscles.
- Dermatomes -- Dermis of skin

Mesoderm

Intermediate ⟶ Urogenital system

Lateral
- Blood vessels and lymphatics (not pharyngeal arch arteries)
- Blood and lymph cells
- Connective tissues
  - coverings of laryngeal muscles

## Cranial Endodermal Derivatives

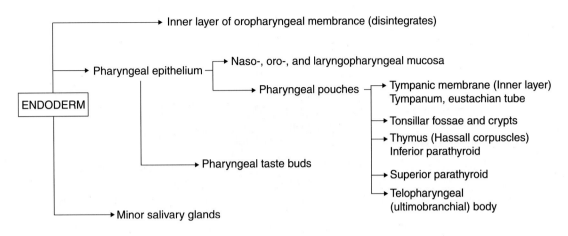

ENDODERM

→ Inner layer of oropharyngeal membrane (disintegrates)

→ Pharyngeal epithelium
- → Naso-, oro-, and laryngopharyngeal mucosa
- → Pharyngeal pouches
  - → Tympanic membrane (Inner layer) Tympanum, eustachian tube
  - → Tonsillar fossae and crypts
  - → Thymus (Hassall corpuscles) Inferior parathyroid
  - → Superior parathyroid
  - → Telopharyngeal (ultimobranchial) body

→ Pharyngeal taste buds

→ Minor salivary glands

## Neural Crest Derivatives

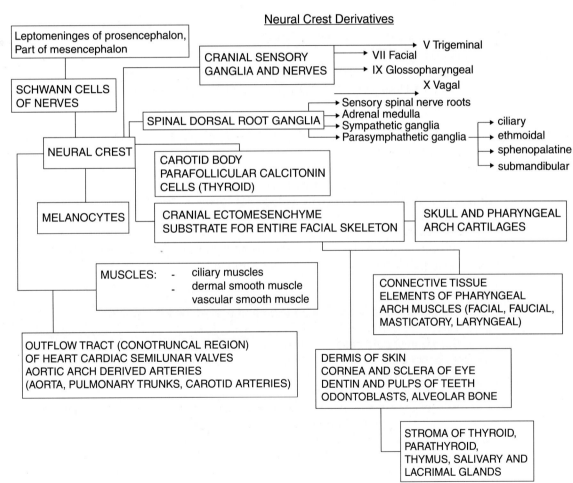

Leptomeninges of prosencephalon, Part of mesencephalon

SCHWANN CELLS OF NERVES

NEURAL CREST

CRANIAL SENSORY GANGLIA AND NERVES
- → V Trigeminal
- → VII Facial
- → IX Glossopharyngeal
- → X Vagal

SPINAL DORSAL ROOT GANGLIA
- → Sensory spinal nerve roots
- → Adrenal medulla
- → Sympathetic ganglia
- → Parasymphathetic ganglia
  - → ciliary
  - → ethmoidal
  - → sphenopalatine
  - → submandibular

CAROTID BODY PARAFOLLICULAR CALCITONIN CELLS (THYROID)

MELANOCYTES

CRANIAL ECTOMESENCHYME SUBSTRATE FOR ENTIRE FACIAL SKELETON

SKULL AND PHARYNGEAL ARCH CARTILAGES

MUSCLES:
- ciliary muscles
- dermal smooth muscle
- vascular smooth muscle

CONNECTIVE TISSUE ELEMENTS OF PHARYNGEAL ARCH MUSCLES (FACIAL, FAUCIAL, MASTICATORY, LARYNGEAL)

OUTFLOW TRACT (CONOTRUNCAL REGION) OF HEART CARDIAC SEMILUNAR VALVES AORTIC ARCH DERIVED ARTERIES (AORTA, PULMONARY TRUNKS, CAROTID ARTERIES)

DERMIS OF SKIN CORNEA AND SCLERA OF EYE DENTIN AND PULPS OF TEETH ODONTOBLASTS, ALVEOLAR BONE

STROMA OF THYROID, PARATHYROID, THYMUS, SALIVARY AND LACRIMAL GLANDS

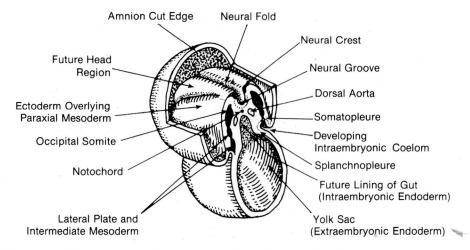

Amnion Cut Edge    Neural Fold

Future Head Region

Ectoderm Overlying Paraxial Mesoderm

Occipital Somite

Notochord

Lateral Plate and Intermediate Mesoderm

Neural Crest

Neural Groove

Dorsal Aorta

Somatopleure

Developing Intraembryonic Coelom

Splanchnopleure

Future Lining of Gut (Intraembryonic Endoderm)

Yolk Sac (Extraembryonic Endoderm)

**Figure 2-13** Transverse section through 20-day-old embryo depicting neural folds and neural crest formation.

## TABLE 2-2: Derivatives of Neural Crest Cells

Connective tissues
    Ectomesenchyme of facial prominences and pharyangeal arches
    Bones and cartilages of facial and visceral skeleton (basicranial and pharyngeal arch cartilages)
    Dermis of face and ventral aspect of neck
    Stroma of salivary, thymus, thyroid, parathyroid and pituitary glands
    Corneal mesenchyme
    Sclera and choroid optic coats
    Blood vessel walls (excepting endothelium); aortic arch arteries
    Dental papilla (dentine, portion of periodontal ligament, cementum)
Muscle tissues
    Ciliary muscles
    Covering connective tissues of pharyngeal arch muscles (masticatory, facial, faucial, laryngeal) combined with mesodermal components
Nervous tissues
    Supporting tissues:
        Leptomeninges of prosencephalon and part of mesencephalon
        Glia
        Schwann sheath cells
    Sensory ganglia:
        Autonomic ganglia
        Spinal dorsal root ganglia
        Sensory ganglia (in part) of trigeminal, facial (geniculate), glossopharyngeal (otic and superior) and vagal (jugular) nerves
    Autonomic nervous system:
        Sympathetic ganglia and plexuses
        Parasympathetic ganglia (ciliary, ethmoid, sphenopalatine, submandibular, enteric system)
Endocrine tissues
    Adrenomedullary cells and adrenergic paraganglia
    Calctonin "C" cells—thyroid gland (ultimopharyngeal body)
    Carotid body
Pigment cells
    Melanocytes in all tissues
    Melanophores of iris

maxillomandibular identity is determined by endothelin-1 (Edn1) *signaling*, perturbation of which leads to fundamental jaw distortion.

Neural crest cells divide as they migrate, forming a larger population at their destination than initially. These cells form the major source of connective tissue components, including cartilage, bone and ligaments of the facial and oral regions, and contribute to the muscles and arteries of these regions.

Cranial neural crest tissue is discontinuous and segments into regions adjacent to the brain. Neural crest cell clusters adjacent to the neural tube form the ganglia of the autonomic nervous system and sensory nerves. From the level of the diencephalon to the midmetencephalon, neural crest tissue provides ectomesenchyme for periocular regions and the first pharyngeal arch; from the myelencephalon to the middle of the otic placode, it migrates to the second pharyngeal arch; that from the midplacodal level to the third somite populates the third to sixth pharyngeal arches, the aortic sac, and parts of the heart (Figure 2–14).

The neural crest cells, being multipotential, display varying regional characteristics at their destination sites: those remaining rostral and dorsal to the forebrain contribute to the leptomeninges and parts of the skull; those around the midbrain form part of the anlagen of the trigeminal nerve ganglion, and in conjunction with the cranial paraxial mesoderm, they form the chondrocranium (see Chapter 8). Interaction with an epithelium (neural or epidermal) is necessary before chondrogenesis can begin.

Neural crest cells migrating ventrally and caudally encounter the pharyngeal endoderm that induces formation of the pharyngeal arches. Many pharyngeal derivatives, including facial bones, are of neural crest origin (see Chapter 4). Those cells that migrate within cranial paraxial mesoderm

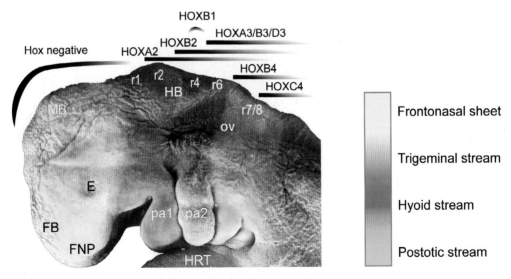

Figure 2–14 A stage 15, 33-day-old human embryo upon which are depicted the neural crest streams emanating from the rhombomeres (r1 –8), influenced by the homeobox (HOX) gene expression patterns. FNP: frontonasal prominence; FB: forebrain; E: eye; MB: midbrain; HB: hindbrain; OV: otic vesicle; HRT: heart; pa1/2: pharyngeal arches 1/2, (SEM *by Prof Steding, Göttingen. By permission of Springer-Verlag.*)

form somitomeres which provide most of the muscles of the face and jaws; other neural crest cells provide mesenchyme for angiogenesis to produce blood vessels; yet others are the source of melanocytes for skin and eye pigmentation.

The developing craniofacial complex is a community of all the above cell populations. Deviation of any one of these populations from its normal development and spatial cueing has deleterious consequences, inducing clusters of anomalies termed *neurocristopathy syndromes*.

### Somite Period

The basic tissue types having developed during the first 21 days, during the next 10 days, development is characterized by foldings and structuring, as well as by differentiation of the basic tissues that convert the flat embryonic disk into a tubular body (Figures 2–15 and to 2–16; also see Figs. 2–12, 2–13, and 2–14). The first of these changes (21 days) is the folding of the neural plate, from which the brain and spinal cord develop. Next, the mesoderm develops into three aggregations—the lateral plate and intermediate and paraxial mesoderm, each with a different fate. The lateral plate mesoderm contributes to the walls of the embryonic coelom from which the pleural, pericardial and peritoneal cavities develop. Lateral mesoderm also forms peripharyngeal connective tissues of the neck. The intermediate mesoderm, absent from the head region, contributes to the formation of the gonad, kidney and adrenal cortex. The paraxial mesoderm, alongside the notochord, forms a rostral condensation of incompletely segmented *somitomeres*. The cranial somitomeric mesoderm migrates into the ventral pharyngeal region to contribute to the future pharyngeal arches (Fig. 2–17). The caudal paraxial mesoderm forms a series of segmental blocks, termed *somites*, whose prominence characterizes this period (21st to 31st days) of embryonic development (Figs. 2–17 and

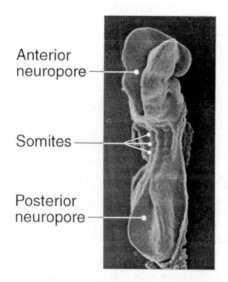

Anterior neuropore

Somites

Posterior neuropore

**Figure 2–15** Scanning electron micrograph of a mouse embryo equivalent to a 22-day-old human embryo with 4 to 12 pairs of somites. The neural folds have fused in the hindbrain region. (Courtesy of Dr. K. K. Sulik, University of North Carolina.)

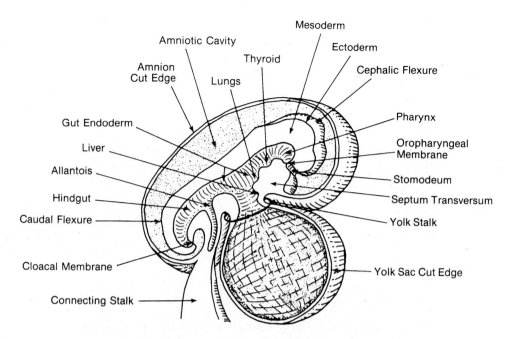

**Figure 2–16** Longitudinal section through a 4-week-old embryo.

2–18). The 42 to 44 paired somites appear sequentially craniocaudally and set the pattern for the regions of the body by their being identified as 4 occipital, 8 cervical, 12 thoracic, 5 lumbar, 5 sacral and 8 to 10 coccygeal somites (Figs. 2–17 and 2–18).

Each somite differentiates into three parts whose fates are implied in their names. The ventromedial part is designated the *sclerotome*\*; it contributes to the vertebral column and accounts for its segmental nature, except in the occipital region, where fusion forms the occipital skull bone. The lateral aspect of the somite, termed the *dermatome*, gives rise to the

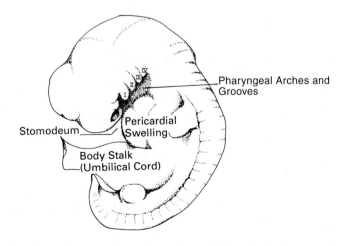

**Figure 2–17** Surface view of late somite period embryo (31 days), showing conspicuous somites along the back and development of the pharyngeal arches and limb buds.

---

\*Proteoglycans (procollagen) and collagen secreted by the notochord induce the conversion of somite sclerotomal cells into cartilage.

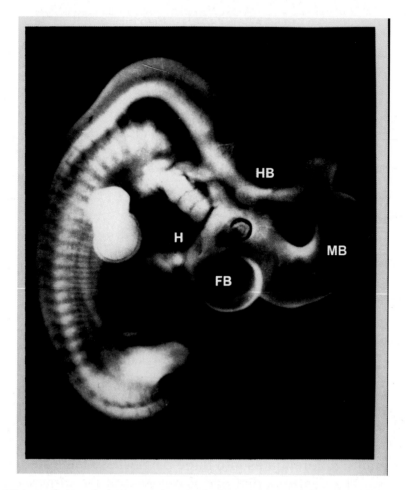

**Figure 2-18** Lateral view of a human embryo in the late somite period (early in the fifth week). Note the forebrain (FB), containing the prominent eye, the midbrain (MB) and hindbrain (HB). The primitive face is flexed in contact with the prominent heart (H). The somites are evident as dorsal segments continuing into a tail. The upper and lower limb buds are seen as prominent paddle-shaped projections. (Courtesy of the late Professor Dr. E. Blechschmidt, University of Göttingen, Germany.)

dermis of the skin. The intermediate portion, the *myotome*, differentiates into muscles of the trunk and limbs and contributes to some of the muscles of the orofacial region.

The somite period is characterized by the establishment of most of the organ systems of the embryo. The cardiovascular, alimentary, respiratory, genitourinary and nervous systems are established and the primordia of the eye and the internal ear appear. The embryonic disk develops lateral, head, and tail folds, facilitating enclosure of the endodermal germ layer from the yolk sac, thereby laying the basis for the tubular intestine. The part of the yolk sac endoderm incorporated into the cranial end of the embryo is termed the *foregut*, the anterior boundary of which is closed off by the oropharyngeal membrane. Similarly, the part of the yolk sac incorporated into the caudal end of the embryo is termed the *hindgut*, bounded ventrally by the cloacal membrane. The intervening portion of the alimentary canal is called the *midgut*, which remains in communication

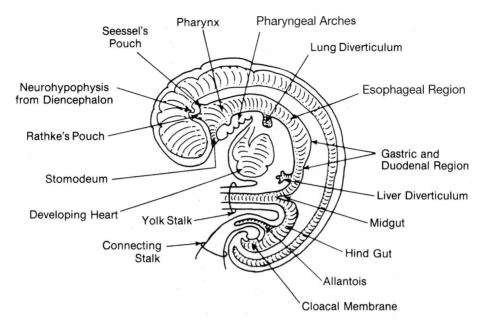

**Figure 2–19** Longitudinal section through a 4.5–week-old embryo depicting the gut and its derivatives.

with the yolk sac through the *yolk stalk* (Fig. 2–16). The gut is initially sealed off at both ends, and is only converted into a canal by the later breakdown of the oropharyngeal and cloacal membranes.

The foregut endoderm later gives rise to the *laryngotracheal diverticulum* from which the bronchi and lungs develop. Other endodermal outgrowths from the foregut are the *hepatic* and *pancreatic diverticula*, giving rise to the secretory elements of the liver and pancreas. The foregut itself develops into the pharynx and its important *pharyngeal pouches*, the oesophagus, stomach and first part of the duodenum. The midgut forms the rest of the duodenum, the entire small intestine, and the ascending and transverse colon of the large intestine. The hindgut forms the descending colon and the terminal parts of the alimentary canal.

The external appearance of the late somite embryo presents a prominent brain forming a predominant portion of the early head, whose face and neck, formed by pharyngeal arches, is strongly flexed over a precocious heart. The eyes, nose and ears are demarcated by placodes, while ventrolateral swellings indicate the beginnings of the limb buds. The lower belly wall protrudes conspicuously in its connections with the placenta through the connecting (body) stalk, and a prominent tail terminates the caudal end of the embryo (see Figs. 2–17 and 2–19).

The extremely complex and very rapid basic organogenesis taking place during the 10-day somite period makes the embryo exceedingly susceptible to environmental disturbances that may produce permanent developmental derangements. Maternal illnesses, particularly of viral origin, irradiation and drug therapy during the first trimester of pregnancy (which includes the somite period of development), are well known in obstetric practice to be the cause of some congenital anomalies of the fetus.

## Postsomite Period

The predominance of the segmental somites as an external feature of the early embryo fades during the 6th week pc. The head dominates much of the development of this period. Facial features become recognizable when the ears, eyes and nose assume a human form, and the neck becomes defined by its elongation and the sheathing of the pharyngeal arches. The paddle-shaped limb buds of the early part of the period expand and differentiate into their divisions to the first demarcation of their digits. The earliest muscular movements are first manifest at this time. The body stalk of the previous periods condenses into a definitive umbilical cord as it becomes less conspicuous on the belly wall. The thoracic region swells enormously, as the heart, which becomes very prominent in the somite period, is joined by the rapidly growing liver, whose size dominates the early abdominal organs. The long tail of the beginning of the embryonic period regresses as the growing buttocks aid in its concealment. The embryo at the end of this period is now termed a *fetus*.

## Fetal Period

The main organs and systems having developed during the embryonic period, the last 7 months of fetal life are devoted to very rapid growth and reproportioning of body components, with little further organogenesis or tissue differentiation. The precocious growth and development of the head in the embryonic period is not maintained in the fetal period, with the result that the body develops relatively more rapidly. The proportions of the head are thereby reduced from about half the overall body length at the beginning of the fetal period to about one-third at the fifth month and about one-fourth at birth. At 4 months pc, the face assumes a human appearance as the laterally directed eyes move to the front of the face and the ears rise from their original mandibulocervical site to eye level. The limbs grow rapidly but disproportionately, the lower limbs more slowly than the upper. Ossification centers make their appearance in most of the bones during this period. The sex of the fetus can be observed externally by the third month, and the wrinkled skin acquires a covering of downy hairs (lanugo) by the fifth month. During this month, fetal movements are first felt by the mother. The sebaceous glands of the skin become very active just before birth (seventh to eighth months), covering the fetus with their fatty secretions, termed the *vernix caseosa* (*vernix*, varnish; *caseosa*, cheesy). Fat makes its first appearance in the face when adipose tissue differentiates and proliferates from the 14th week pc onward. It appears initially in the buccal fat pad area, then in the cheek, and finally in the chin subcutis. In the last 2 months of fetal life, fat is deposited subcutaneously to fill out the wrinkled skin, and nearly half the ultimate birth weight is added.

Despite the rapid growth of the postcranial portions of the body during the fetal period, the head still has the largest circumference of all the parts of the body at birth. The passage of the head through the birth canal has to be accommodated by its compression. The birth compression of the cranium presents the danger of distortion; this normally rectifies itself postnatally, but may persist as a source of perverted mandibulofacial development.

### Anomalies of Development

The embryonic prechordal plate is a significant median suppressor of the rostral end of the neural plate that accordingly is subdivided into bilateral components to form the two telencephalic hemispheres. Lack of SHH and PAX6 *signaling* from the prechordal plate results in an undivided forebrain, known as holoprosencephaly, creating varyingly severe facial anomalies, ranging from cyclopia to a single median maxillary central incisor tooth.

Neural tube defects (NTDs), leading to a spectrum of congenital anomalies, varying from mere disfigurement to lethal conditions, are among the more common congenital anomalies. Such defects are attributed to, among others, deficiencies of follistatin, a maternally expressed protein, whose production is dependent on nutritional folic acid, a water-soluble vitamin. Adequate folic acid intake during early pregnancy minimizes the incidence of NTDs. A difference exists between male and female embryos during neural fold elevation due to the presence in female cells of an inactivated highly methylated second X chromosome. Methylation is essential to neural fold elevation, resulting in a higher incidence of NTDs in females.

Neural tube closure occurs at several sites (Fig. 2–20). Closure is initiated at the boundary between the future hindbrain and the spinal cord (site 1) and at further three distinct sites in the cranial region (sites 2, 3 and 4). A fifth closure site, involving a process of canalization, occurs in the caudal region (site 5). The directions of closure are indicated by the arrows. Failure of closure results in neural tube defects (right hand panels). The top row depicts anencephaly and craniorachischisis. Encephaloceles

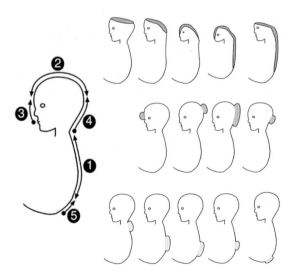

**Figure 2–20** Schematic depiction of neural tube defects. Sites of sequential neural tube closure points (1–5) and congenital anomalies consequent to failure of closure. (Courtesy of Dr. D. Alan Underhill, University of Alberta.)

(middle row) involve failures to complete neural tube closure or membrane fusion at closure points (sites 2, 3 and 4) within the cranial region. Spina bifida cystica (bottom row) comprises a range of neural tube defects of the spinal cord and its coverings at various levels (sites 1 and 5).

## SELECTED BIBLIOGRAPHY

Anderson DJ. Cellular and molecular biology of neural crest cell lineage determination. Trends Genet 1997; 13:276–280.

Aoto K, Shikata Y, Imai H, et al. Mouse Shh is required for prechordal plate maintenance during brain and craniofacial morphogenesis. Dev Biol 2009; 327:106–120.

Ezin AM, Fraser SE, Bronner-Fraser M. Fate map and morphogenesis of presumptive neural crest and dorsal neural tube. Dev Biol 2009; 330:221–236.

Hirokawa N, Tanaka Y, Okada Y. Left-right determination: involvement of molecular motor KIF3, cilia, and nodal flow. Cold Spring Harbor Perspect Biol 2009; a000802; 1–22.

Johnson MC, Bronsky PT. Prenatal craniofacial development: new insights on normal and abnormal mechanisms. Crit Rev Oral Biol Med 1995; 6:368–422.

Killion ECD, Birkholz DA, Artinger KB. A role for chemokine signaling in neural crest migration and craniofacial development. Dev Biol 2009; 333: 161–172.

Kjaer I. Human prenatal craniofacial development related to brain development under normal and pathologic conditions. Acta Odont Scand 1995; 53:135–143.

La Bonne C, Bronner-Fraser M. Molecular mechanisms of neural crest formation. Annu Rev Cell Dev Biol 1999; 15:81–112.

Lallier TE. Cell lineage and cell migration in the neural crest. Ann NY Acad Sci 1991; 615:158–171.

LeDouarin NM, Calloni GW, Dupin E. The stem cells of the neural crest. Cell Cycle 2008; 7:1013–1019.

Lee VM. Developmental potential of migrating neural crest cells. Dev Biol 2008; 319:533.

Lee JD, Anderson KV. Morphogenesis of the node and notochord: the cellular basis for the establishment and maintenance of left-right asymmetry in the mouse. Dev Dyn 2008; 237: 3464–3476.

Muller F, O'Rahilly R. The prechordal plate, the rostral end of the notochord and nearby median features in staged human embryos. Cell Tiss Org 2003; 173:1–20.

Sato T, Kurihara Y, Sai R, et al. An endothelin-1 switch specifies maxillomandibular identity. PNAS 2008; 105:18806–18811.

Steventon B, Araya C, Linker C, et al. Differential requirements of BMP and Wnt *signaling* during gastrulation and neurulation define two steps in neural induction. Development 2009: 136: 771–779.

Tanaka O. Variabilities in prenatal development of orofacial system. Anat Anz 1991; 172:97–107.

Yoshida T, Vivatbutsiri P, Morriss-Kay G, et al. Cell lineage in mammalian craniofacial mesenchyme. Mech Dev 2008; 125: 797–808.

Zamir EA, Rongish BJ, Little CD. The ECM moves during primitive streak formation— computation of ECM versus cellular motion. PLoS Biol 2008; 6(10): e247. doi:10.1371/journal.pbio.0060247

# 3 Early Orofacial Development

Development of the head depends upon the prior presence of the brain, whose rostral parts, the prosencephalon and the mesencephalon, are specified by the ortodenticle homologues of OTX1 and OTX2, while overlapping HOX genes specify gene expression, in a rostrocaudal manner, the rhombencephalon (rhombomeres 2 to 7), and its neural crest derivatives. The inductive *signaling* activities that emanate from the prosencephalic and rhombencephalic organizing centers are regulated by the Sonic hedgehog (SHH) gene that is expressed as a signaling protein (Fig. 3–1).

The *prosencephalic center*, derived from prechordal mesoderm that migrates from the primitive streak, is at the rostral end of the notochord beneath the forebrain (prosencephalon); it induces the visual and inner ear apparatuses and upper third of the face. The caudal *rhombencephalic center* induces the middle and lower thirds of the face (the viscerofacial skeleton), including the middle and external ears.

Associated with these developments is the division of the initially unilobar forebrain (Fig. 3–2) into paired telencephalic hemispheres and evagination of the paired olfactory bulbs and optic nerves. Failure of these cerebral divisions (holoprosencephaly) has a profound influence on facial development, leading to many types of anomalies.

The forebrain establishes multiple signaling centers in the ectoderm that covers the nascent frontonasal zone. This brain SHH signaling center imprints on the forebrain ectoderm, controlling its differential cell proliferation (growth zones) that transforms the undifferentiated frontonasal prominence into the characteristic sculpted undulating nasal regions of

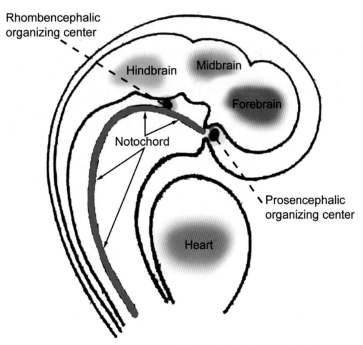

**Figure 3–1** Schematic depiction of prosencephalic and rhombencephalic organizing centers.

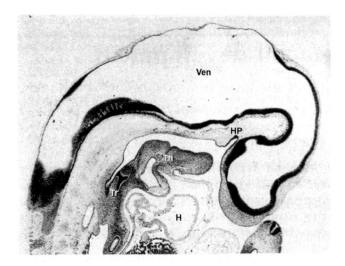

**Figure 3–2** Sagittal section of head region of a 32-day-old embryo. The ventricles (Ven) occupy a large proportion of the brain. The hypophyseal pouch (HP) extends up from the stomodeum, which is bordered below by the heart (H). The thyroid gland (Th) is budding from the foramen cecum in the tongue via the thyroglossal duct. The trachea (Tr) is budding from the primitive esophagus (× 22, reduced to 70% on reproduction). (Courtesy of Professor H. Nishimura.)

the upper face. Subtle changes in the gene expression patterns in the frontonasal ectodermal growth zone may result in facial malformations varying from mild to extreme.

Oral development in the embryo is demarcated extremely early in life by the appearance of the prechordal plate in the bilaminar germ disk on the 14th day of development, even before the mesodermal germ layer appears. The endodermal thickening of the prechordal plate designates the cranial pole of the oval embryonic disk, and later contributes to the *oropharyngeal membrane*. This tenuous and temporary bilaminar membrane is the site of junction of the ectoderm that forms the mucosa of the mouth and the endoderm that forms the mucosa of the pharynx, which is the most cranial part of the foregut. The oropharyngeal membrane is one of two sites of contiguity between ectoderm and endoderm, where mesoderm fails to intervene between the two primary germ layers; the other site is the cloacal membrane at the terminal end of the hindgut. The oropharyngeal membrane also demarcates the site of a shallow depression, the *stomodeum*, the primitive mouth that forms the topographical center of the developing face. Rapid orofacial development is characteristic of the more advanced development of the cranial portion of the embryo than its caudal portion. The differential rates of growth result in a pear-shaped embryonic disk, the head region forming the expanded portion of the pear (see Fig. 2–10). Further, the three germ layers in the cranial part of the embryo begin their specific development by the middle of the third week, whereas separation of the germ layers continues in the caudal portion until the end of the fourth week pc. Because of the precocious development of the cranial end of the embryo, the head constitutes nearly half the total body size during the postsomite embryonic period (fifth to eighth weeks); with subsequent postcranial development, it forms only one-quarter of the body length at birth, but 6% to 8% of the body in adulthood.

## FORMATION OF THE FACE

### Genetics

The tumor protein 63 transcription factor (TP63) gene is expressed in proliferation and differentiation of the basal cells of epithelial tissues that provide the foundation of development of the orofacial ensemble. The interaction of this gene upon BMP, FGF8 and SHH *signaling* controls the growth, modeling and fusion events underlying facial formation.

The face derives from five prominences that surround a central depression, the stomodeum, which constitutes the future mouth. The prominences are the single median *frontonasal* and the paired *maxillary* and *mandibular prominences* (Figs. 3–3 and 3–4); the latter two are derivatives of the first pair of six pharyngeal arches. All of these prominences and arches arise from neural crest ectomesenchyme which migrates from its initial dorsal location into the facial and neck regions.

The frontonasal prominence surrounds the forebrain, which sprouts lateral optic diverticula that form the eyes. The frontal portion of the prominence between the eyes forms the forehead; at the inferolateral corners, thickened ectodermal nasal (olfactory) placodes arise (Figs. 3–5, 3–5A).

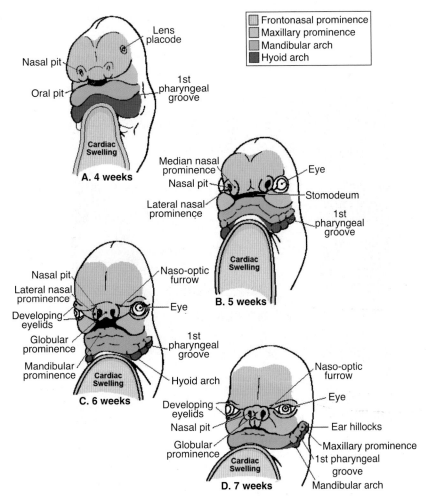

**Figure 3–3** Schematic depiction of facial formation: A: 4 weeks; B: 5 weeks; C: 6 weeks; D: 7 weeks.

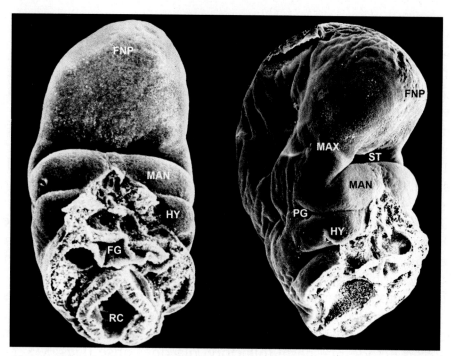

**Figure 3–4**  Scanning electron micrographs of the head region of a 26-day-old human embryo. The central stomodeum (ST) is bordered by the frontonasal prominence (FNP) above, the maxillary (MAX) prominences laterally and the mandibular (MAN) prominences inferiorly. The hyoid (HY) arch borders the first pharyngeal groove (PG). The sectioned surface reveals the foregut (FG) and the rhombencephalon (RC). (Scale bar = 0.1 mm) (From Hinrichsen K. The early development of morphology and patterns of the face in the human embryo. In Advances in anatomy, embryology and cell biology. New York: Springer Verlag, 1985:98; by permission.)

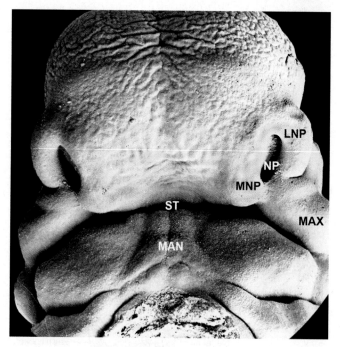

**Figure 3–5**  Scanning electron micrograph of the face of a 37-day-old human embryo. The nasal pits (NP) are surrounded by the medial (MNP) and lateral (LNP) nasal prominences and the maxillary (MAX) prominences. The wide stomodeum (ST) is limited inferiorly by the mandibular prominence (MAN). (From Hinrichsen K. The early development of morphology and patterns of the face in the human embryo. In Advances in anatomy, embryology and cell biology. New York: Springer Verlag, 1985:98; by permission; see Figure 3–4.)

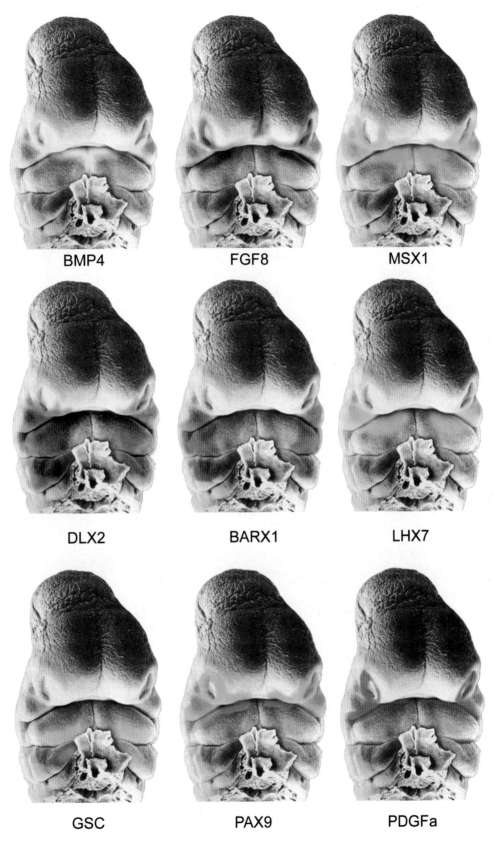

BMP4     FGF8     MSX1

DLX2     BARX1     LHX7

GSC     PAX9     PDGFa

**Figure 3–5A** Scanning electron micrographs of the face of a Stage 15, 33-day-old human embryo depicting the gene expression patterns derived from mouse embryos. (Faces from Hinrichsen; by kind permission of Springer-Verlag.)

These placodes, resulting from the stepwise combinatorial expression of DLX3, DLX5, PAX6 and other transcription factors, develop olfactory epithelium and olfactory neurons that form the underlying olfactory nerves. The olfactory placodes become invaginated by the elevation of inverted horseshoe-shaped ridges, the *medial* and *lateral nasal prominences*, which surround each sinking *nasal pit* (Fig. 3–6). These pits are precursors of the anterior nares, initially in continuity with the stomodeum.

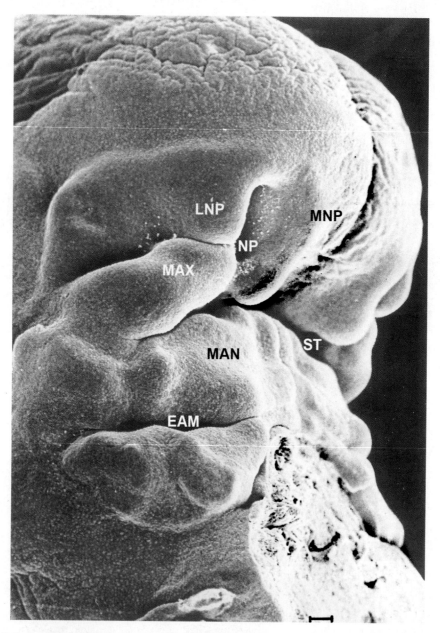

**Figure 3–6** Scanning electron micrograph of a 41-day-old human embryo. The maxillary prominence (MAX) is wedged between the lateral (LNP) and medial (MNP) nasal prominences surrounding the nasal pit (NP). The stomodeum (ST) is bordered inferiorly by the mandibular (MAN) prominence. Auricular hillocks are forming around the external acoustic meatus (EAM). (Scale bar = 0.1 mm) (From Hinrichsen K. The early development of morphology and patterns of the face in the human embryo. In Advances in anatomy, embryology and cell biology. New York: Springer Verlag, 1985:98; by permission; see Figure 3–4.)

The nasal placodes provide morphogenetic information to the adjacent lateral nasal prominence mesenchyme, acting as a signaling center for the lateral nasal skeleton, inducing cartilage and bone. The nasal placodes are the source of the olfactory epithelium and olfactory neurons. Union of the facial prominences occurs by either of two developmental events at different locations by *merging* of the frontonasal, maxillary and mandibular prominences, or by *fusion* of the central maxillonasal components. Merging of what are, initially, incompletely separated prominences occurs as the intervening grooves disappear as a result of migration into and/or proliferation of underlying mesenchyme in the groove. Fusion of the freely projecting medial nasal prominences with the maxillary and lateral nasal prominences on each side requires disintegration of their contacting surface epithelia (the *nasal fin*), allowing intermingling of the underlying mesenchymal cells, consequent to cell proliferation, extracellular matrix production, vascular invasion and fluid accumulation. Upper lip formation commences at 24 days postconception and is completed by 37 days. Formation of the upper lip is a complex process involving WNT Lrp6, SHH, FGF and BMP *signaling* pathways that pattern cell proliferation and tissue configuration. Failure of normal disintegration of the nasal fin, by cell death or epithelial-mesenchymal transformation, is a cause of cleft upper lip and the anterior palate, by preventing the merging of maxillary and medial nasal mesenchyme.

Fusion of the medial nasal and maxillary prominences provides for continuity of the upper jaw and lip and separation of the nasal pits from the stomodeum. Although the lateral nasal prominences do not contribute to the upper lip, failure of their prior initial fusion with the medial nasal prominences is a factor causing clefts of the upper lip that extend into the nostril. The midline merging of the medial nasal prominences forms the median tuberculum and philtrum of the upper lip, the tip of the nose and the primary palate. The intermaxillary segment of the upper jaw (the "premaxilla"), in which the four upper incisor teeth will develop, arises from the median primary palate that initially is a widely separated pair of inwardly projecting swellings of the merged medial nasal prominences (Fig. 3–7). Signaling through transforming growth factor-$\beta 1$ (TGF-$\beta 1$) receptor ALK5, SHH, FGF and BMP that pattern cell proliferation and tissue configuration is required for upper lip fusion. Abnormal bilateral clefting resulting from failure of fusion of medial nasal and maxillary prominences produces a prominent projection of the merged medial nasal prominences (globular process) (see Fig. 3–23).

The lower jaw and lip are very simply formed by midline merging of the paired mandibular prominences, and are the first parts of the face to become definitively established. The lateral merging of the maxillary and mandibular prominences creates the commissures (corners) of the mouth.

During the seventh week pc, a shift in the blood supply of the face, from the internal to the external carotid artery, occurs as a result of normal atrophy of the stapedial artery (Fig. 3–8). This shift occurs at a critical time

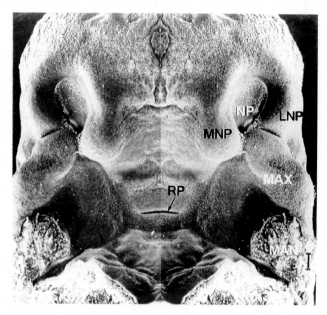

**Figure 3–7** Scanning electron micrograph of the stomodeal chamber of a 41-day-old human embryo. The maxillary prominences (MAX) are wedged between the lateral (LNP) and medial (MNP) nasal prominences surrounding the nasal pits (NP). The mandibular prominences (MAN) are cut. Rathke's pouch (RP) is in the roof of the stomodeum. (Scale bar = 0.1 mm) (From Hinrichsen K. The early development of morphology and patterns of the face in the human embryo. In Advances in anatomy, embryology and cell biology. New York: Springer Verlag, 1985:98; by permission; see Figure 3–4.)

of midface and palate development, providing the potential for deficient blood supply and consequent defects of the upper lip and palate.

Not all regions of the face grow equally rapidly during early development. There is relative constancy of the central facial region (between the eyes) in relation to the rapidly expanding lateral facial regions and a reduction of interocular distance. These changes confer human characteristics upon the developing face (Fig. 3–9). Malproportioning of growth in this time is the basis of developing craniofacial defects.

### The Eyes

A median single field of cells—the optic primordium—in the anterior neural plate forming the ventral anterior diencephalon will develop bilateral retinas under the direction of the *cyclops* gene. Cyclopic *signaling* induces divergent morphogenetic movements of the ventral diencephalon to form bilateral eyes. The prechordal plate suppresses the median part of the optic primordium to form bilateral primordia. The ventral anterior diencephalon fails to form in the absence or mutation of the *cyclops* gene, resulting in a single cyclopic eye.

Lateral expansions of initial forebrain evaginations (optic sulci) form optic vesicles, which medially retain their diencephalic (forebrain) connections (the optic stalks) and laterally induce thickened epithelial *lens placodes* on the sides of the future face (Fig. 3–10). Invagination of the lens placode concomitantly with formation of optic vesicles creates the deep-set eyeballs (see Chapter 18). Medial migration of the eyes from their

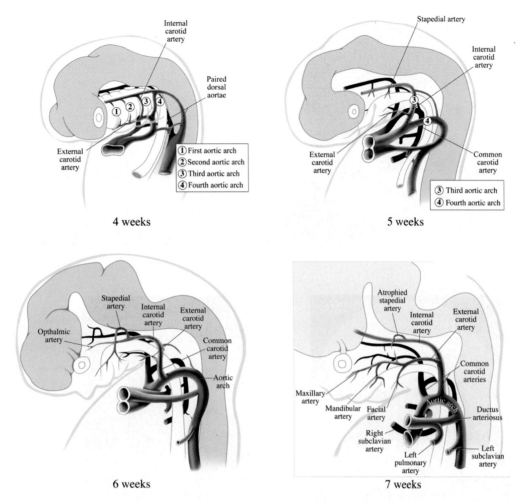

**Figure 3–8** Arterial blood supply to the head at various ages. Atrophy of the stapedial artery shifts the facial blood supply from the internal to the external carotid artery.

initial lateral locations results from enormous growth of the cerebral hemispheres and broadening of the head as well as real medial movement of the eyes. The greatest migratory movements of the eyes occur from the fifth to ninth weeks pc; thereafter they stabilize to their postnatal angulation of the optic axes, at 71° to 68°.

Early in fetal life (8 weeks), folds of surface ectoderm overgrow the eyes to form the eyelids (see Fig. 3–9 [c and d]). These remain fused until the seventh month pc, and then invading muscle allows their opening.

## The Ears

The internal ear first manifests as a hindbrain induction of surface ectodermal cells elongating into a thickened otic placode. The placode later invaginates into a pit, which closes off as a vesicle and thus forms the internal ear (see Chapter 18).

The external ear develops in the neck region as six auricular hillocks surrounding the first pharyngeal groove that forms the external acoustic meatus (see Figs.3–6 and 3–10; Fig. 3–11). This combination of elevations

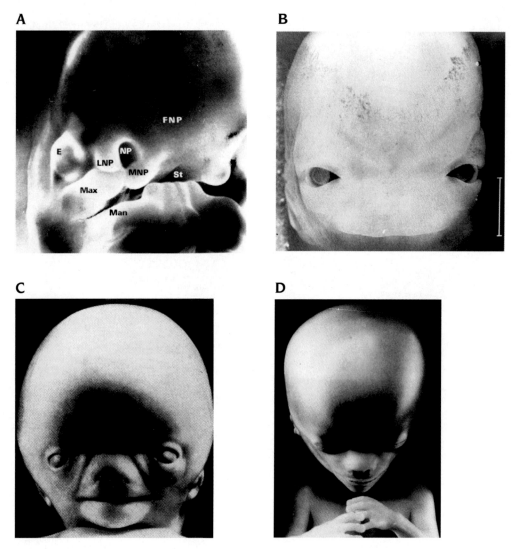

**Figure 3–9** (A) Face of a human embryo in a late somite period (30–32 days). The frontonasal prominence (FNP) projects the medial nasal (MNP) and lateral nasal (LNP) prominences on each side, surrounding the nasal placode. The eye (E) is seen in its lateral position. The maxillary prominence (Max) forms the superolateral boundary of the stomodeum (St) and the mandibular prominence (Man) forms the lower boundary. (B) Photomicrograph of a 54-day-old human embryo depicting widely separated eyes prior to eyelid closure. Note superficial blood vessels in forehead. (Scale bar = 1 mm) (C) Face of a 7-week-old human embryo. The relatively large forehead dominates the face. Eyelids are beginning to form above and below the migrating eyes. The mouth opening is becoming smaller. (D) Face of a 3-month-old fetus. The eyelids, their formation complete, are closed over the eyes. Medial migration of the eyes will narrow the interocular distance. Precocious brain development dominates the size of the head at this age. Ossification has started to form the craniofacial skeleton. (a, c and d Courtesy of the late Professor Dr. E. Blechschmidt; b from Hinrichsen K. The early development of morphology and patterns of the face in the human embryo. In Advances in anatomy, embryology and cell biology. New York: Springer Verlag, 1985:98; by permission; see Figure 3–4.)

around a depression forming the auricle rises up the side of the developing face to its ultimate location.

The middle ear has a complex origin from the first pharyngeal pouch. Full details of its development are given in Chapter 18.

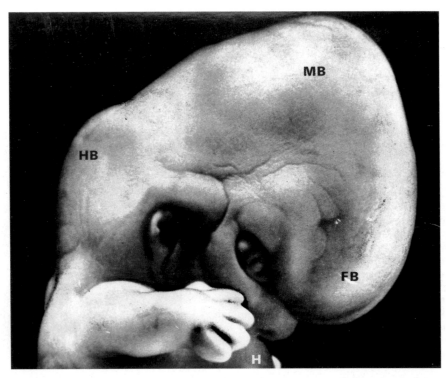

**Figure 3–10** Lateral view of a 6 1/2-week-old embryo. The face is strongly flexed over the prominent heart (H). Note the location of the forebrain (FB), midbrain (MB) and hindbrain (HB) and the first pharyngeal groove forming the external acoustic meatus. The fingers are differentiating out of the hand paddle. Note the absence of eyelids and the prominent lens vesicle in the center of the optic cup. (Courtesy of the late Professor Dr. E. Blechschmidt.)

## The Nose

The nose is a complex of contributions from the frontal prominence (the bridge), the merged medial nasal prominences (the median ridge and tip), the lateral nasal prominences (the alae) and the cartilage nasal capsule (the septum and nasal conchae). The internal and external nasal regions develop from two distinct morphogenetic fields: the deep capsular field gives rise to the cartilaginous nasal capsule and its derivatives; the superficial alar field gives rise to the external alar cartilages.

The nasal pits, previously described, become separated anteriorly from the stomodeum by fusion of the medial nasal, maxillary and lateral nasal prominences to form the nostrils (anterior nares) (see Fig. 3–11). The blind sacs of the deepening nasal pits are initially separated from the stomodeum by the *oronasal membranes* that upon disintegration establish the primitive posterior nares (primary choanae) (Figs. 3–12 and 3–13). The definitive choanae of the adult are created by fusion of the secondary palatal shelves (described below). While the nostrils become patent early in fetal development, exuberant epithelial growth fills them with plugs until midfetal life.

Within the frontonasal prominence a mesenchymal condensation forming the precartilaginous *nasal capsule* develops around the primary nasal cavities as a median mass—the *mesethmoid*, a prologue to the nasal

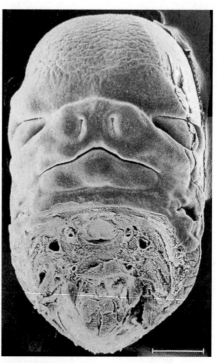

Figure 3–11 Scanning electron micrograph of the face of a 44-day-old human embryo. The merging facial prominences have eliminated the intervening furrows, and diminished the oral opening. The cut surface reveals arteries, veins, the pharynx and the spinal cord. Auricular hillocks are forming around the external acoustic meatus. (Scale bar = 0.1 mm) (From Hinrichsen K. The early development of morphology and patterns of the face in the human embryo. In Advances in anatomy, embryology and cell biology. New York: Springer Verlag, 1985:98; by permission; see Figure 3–4.)

septum—and paired lateral masses, the *ectethmoid*, that will form the paired ethmoidal (conchal) and nasal alar cartilages (Fig. 3–15). A cartilaginous nasal floor does not develop in the human. Initially, the primitive nasal septum is a broad area between the primary choanae, and never projects as a free process, but builds up in a rostrocaudal direction as the later-developing palatal shelves fuse. Within the nasal septum, cartilage forms in continuity with the mesethmoid cartilage, a component of the early basicranium. The nasal septum divides the nasal chamber into left and right fossae; their lateral walls, derived from the ectethmoid, are subdivided into superior, middle and inferior conchae. The nasal mucosa folds, forming the conchae; later these are invaded by cartilages (Fig. 3–15).

Invasion of ectoderm into the median nasal septum from both nasal fossae between the sixth and eighth weeks forms the bilateral *vomeronasal organs* (of Jacobson), vestigial chemosensory structures (Fig. 3–15). The organs form blind pouches, reaching their fullest development at the fifth month pc. Thereafter, they diminish and usually disappear, but may persist as anomalous cysts.

### Nasolacrimal Duct

Within the grooves between the lateral nasal and maxillary prominences, solid rods of epithelial cells sink into the subjacent mesenchyme. The bilateral rods, which extend from the developing conjunctival sacs at the medial corners of the forming eyelids to the external nares, later canalize to form the *nasolacrimal sacs* and *ducts*. The ducts become completely patent only after birth.

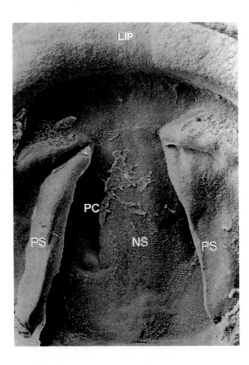

Figure 3–12 Scanning electron micrograph of the stomodeal chamber of a 56-day-old human embryo revealing the nasal septum (NS) and primary choanae (PC) in the nasal chamber. The palatal shelves (PS) have not yet fused. The upper lip (LIP) is fully fused, although the median primary palate has not yet appeared. (From Hinrichsen K. The early development of morphology and patterns of the face in the human embryo. In Advances in anatomy, embryology and cell biology. New York: Springer Verlag, 1985:98; by permission; see Figure 3–4.)

## Early Palate Formation

The primitive stomodeum that forms a wide central shallow depression in the face is limited in its depth by the *oropharyngeal membrane*. The characteristically deep oral cavity is formed by ventral growth of the prominences surrounding the stomodeum. The stomodeum establishes as an oronasopharyngeal chamber and entrance to the gut on the 28th day, when the dividing oropharyngeal membrane disintegrates, providing continuity of passage between the mouth and pharynx.*

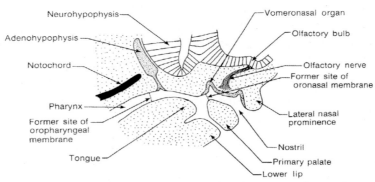

Figure 3–13 Sagittal section through anterior craniofacial region in a late embryo (approximately 7 weeks).

*The oropharyngeal membrane demarcates the junction of ectoderm and endoderm in the embryo. It is very difficult to trace this line of division subsequently because of the extensive changes occurring in oropharyngeal development. As the hypophysis originates from Rathke's pouch anterior to the oropharyngeal membrane, a measure of the depth of oral growth is provided by its adult location. The presumed site of the original oropharyngeal membrane in the adult is an imaginary oblique plane from the posterior border of the body of the sphenoid bone through the tonsillar region of the fauces to the sulcus terminalis of the tongue.

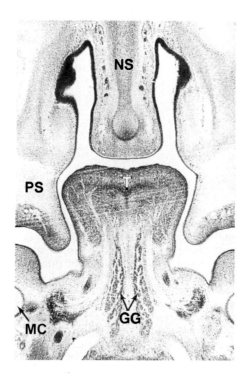

**Figure 3–14** Photomicrograph of coronal section of the stomodeum of a 52-day-old human embryo depicting the vertically oriented palatal shelves (PS) on each side of the tongue (T). Note the nasal septum (NS), genioglossus muscles (GG) and Meckel's cartilage (MC). (From O'Rahilly, R. A colour atlas of embryology. Philadelphia: W.B. Saunders, 1975; by permission.)

The stomodeal chamber divides into separate oral and nasal cavities when the frontonasal and maxillary prominences develop horizontal extensions into the chamber. These extensions form both the central part of the upper lip (the tuberculum) and the single median primary palate from the frontonasal prominence and the two lateral palatal shelves from the maxillary prominences (see Fig. 3–12; Fig. 3–16). The coincidental development of the tongue from the floor of the mouth fills the oronasal chamber, intervening between the lateral palatal shelves; these shelves are vertically orientated initially, but become horizontal when the stomodeum expands and the intervening tongue descends (see Fig. 3–14). The shelves elevate unevenly with the anterior third "flipping up," followed by an oozing "flow" of the posterior two-thirds.

Elevation of the shelves enables their mutual contact in the midline, and with the primary palate anteriorly and the nasal septum superiorly (see Fig. 3–15). Fusion of the shelves, which starts a third of the way from the front, proceeds both anteriorly and posteriorly. The shelves also fuse with the nasal septum, except posteriorly, where the soft palate and uvula remain unattached.

Ossification provides the basis for the anterior bony hard palate. The posterior third of the palate remains unossified: mesenchyme migrates into this region from the first and fourth pharyngeal arches to provide the soft palate muscles, which retain their initial innervation. This process is detailed in Chapter 10.

## Adenohypophyseal Pouch

Ectodermal epithelium invaginates upward in the roof of the primitive stomodeum, immediately ventral to the oropharyngeal membrane,

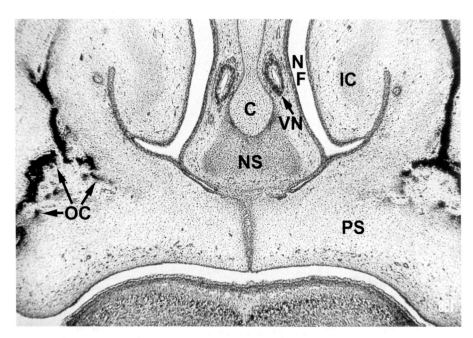

**Figure 3–15** Photomicrograph of coronal section of the palate of a 57-day-old human embryo depicting fusion of the palatal shelves (PS) with each other and the nasal septum (NS). Epithelial degeneration is seen at the fusion lines. The cartilage (C) of the nasal septum separates the vomeronasal organs (VN). Primary ossification centers (OC) of the maxillae are evident. The inferior conchae (IC) are forming from the ectethmoid. The tongue (T) is separated from the nasal fossae (NF). (From O'Rahilly, R. A colour atlas of embryology. Philadelphia: W.B. Saunders, 1975; by permission.)

to form the *adenohypophyseal* (*Rathke's*) *pouch* and *duct* (see Fig. 3–7). The location of this diverticulum coincides with the cranial termination of the underlying notochord and the prechordal mesenchyme that is responsible for inducing the diverticulum. This pouch gives rise to the anterior lobe of the pituitary gland (adenohypophysis), by differentiation of the stomodeal ectoderm into endocrine cells; a downward diverticulum of the diencephalon of the forebrain gives rise to the pituitary's posterior lobe (neurohypophysis). Invading neural crest tissue forms the connective tissue stroma and endocrine cells that produce adrenocorticotropic and melanocyte-stimulating hormones.

The adenohypophyseal pouch normally loses contact with the oral ectoderm by atrophy of the adenohypophyseal duct, but remnants may persist within the sphenoid bone as the *craniopharyngeal canal*. These remnants may become cysts or tumors within the body of the sphenoid bone. Persistence of adenohypophyseal tissue at the site of origin of the hypophyseal duct forms the rare *pharyngeal hypophysis*.

### Seessel's Pouch

A small endodermal diverticulum develops from the cranial end of the foregut, known as Seessel's pouch (see Fig. 2–19). It arises close to the prechordal plate and projects toward the brain. Seessel's pouch may be represented in the adult by the pharyngeal bursa, a depression in the nasopharyngeal tonsil.

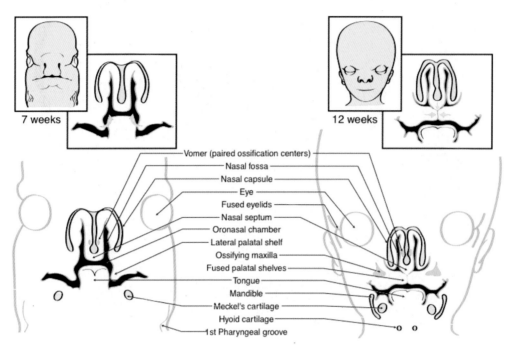

**Figure 3–16** Nasal capsule and palate development in coronal section at 7 weeks (a) and 12 weeks (b) pc.

## Oral Cavity

The oral cavity and entire intestinal tract are sterile at birth. As soon as feeding by mouth commences, an oral bacterial flora is established that will form part of the oral environment throughout life. The oral soft tissues develop a local resistance to infection by this flora which becomes characteristic of the individual mouth.

The facial contours are expanded in the later stages of fetal development by the formation of a subcutaneous panniculus adiposus, minimal in the scalp and thickest in the cheek. This buccal fat pad minimizes collapse of the cheeks during suckling. The pale skin of the neonate darkens with melanocyte activity after birth.

## DEVELOPMENT OF THE CRANIAL NERVES

Initially, the 12 cranial nerves are organized segmentally and numbered sequentially. This serial pattern is not evident in adulthood because of the migrations of the nerves' target organs. The tortuous adult nerve pathways trace the embryonic routes of travel of the innervated organs (Fig. 3–17).

The olfactory nerves arise from bipolar neurons that ingress from the olfactory placodes to synapse with the olfactory bulb of the forebrain. Olfactory neurons are uniquely replaced approximately every 40 days by neural stem cells, exemplifying postnatal neurogenesis in the central nervous system.

The trigeminal (V), facial (VII) and glossopharyngeal (IX) ganglia, and parts of the vestibulocochlear (VIII) and vagal (X) nerve ganglia, derive from the neural crest. Neuroblasts derived from superficial ectodermal

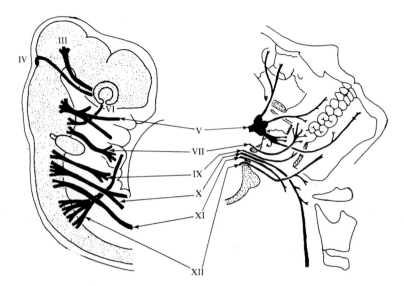

**Figure 3–17** Distribution of branches of the cranial nerves in the fetus and adult. III, Oculomotor nerve; IV, trochlear nerve; V, trigeminal nerve; VI abducens nerve; VII, facial nerve; IX, glossopharyngeal nerve; X, vagus nerve; XI, spinal accessory nerve; XII, hypoglossal nerve.

placodes in the maxillomandibular region contribute to the ganglia of the trigeminal, facial, vagus and glossopharyngeal nerves. The surface placodal contributions to these latter nerves may account for the gustatory components. The vestibulocochlear ganglion arises from the otic vesicle, neural crest cells and an ectodermal placode. Caudal to the vagal ganglion, neural crest tissue of the occipital region contributes to ganglia of the accessory (XI) and hypoglossal (XII) nerves.

The central processes of the neurons of these ganglia form the afferent sensory roots of the cranial nerves. Their peripheral processes form the efferent motor components of the nerves.

Sensory innervation of the face by the trigeminal nerve is based upon its embryological origins. The frontonasal prominence is innervated by the ophthalmic division, the maxillary prominence by the maxillary division, and the mandibular prominence by the mandibular division of the trigeminal nerve; this accounts for the adult pattern of facial sensory innervation, the densest of any cutaneous field. The motor nerve supply is primarily by the facial nerve. The facial nerve contains axons from both branchiomotor and visceromotor neurons. The branchiomotor neurons arise from rhombomere 6 to pass through the geniculate ganglion to innervate facial muscles. The facial visceromotor neurons arise from rhombomere 6 to innervate the submandibular ganglion as the chorda tympani nerve, and the sphenopalatine ganglion as the greater superficial petrosal nerve. For details see Chapter 17.

## ANOMALIES OF DEVELOPMENT

Craniofacial nerve anomalies include the Moebius syndrome, which results in congenital palsy and nerve weakness causing facial asymmetry, due to defects of the abducens (VI), and facial (VII) nerves. Contrastingly,

Bell's palsy is an acquired viral infection of the facial nerve that leads to facial asymmetry.

Defects of facial development are the result of a multiplicity of etiological factors, some genetic, most unknown. The study of these anomalies constitutes *teratology*.

Defective development is categorized into *malformations* that are generally genetically determined, *deformations* that are environmentally influenced and *disruptions* that are of metabolic, vascular and/or teratogenic origin. Malformations develop predominantly in the embryonic period and are not self-correcting, whereas deformations and disruptions occur in the fetal period and may correct themselves (see Fig. 2–1).

The range of facial anomalies is enormous, but all produce some degree of disfigurement and result in impairment of some degree of function or even incompatibility with life. The mechanisms of facial maldevelopment are ill understood, but inductive phenomena arising from the cerebral prosencephalic and rhombencephalic organizing centers are essential to normal facial development. Defective brain development and distorted signaling derived from the brain almost inevitably causes cranial or facial dysmorphism.

*Acephaly*, absence of the head, is the most extreme defect. Postcranial structures can continue developing in utero; however, the condition is lethal upon birth. Absence of the brain (anencephaly) results in *acrania* (absent skull), *acalvaria* (roofless skull) or *cranioschisis* (fissured cranium), with variable effects upon the face (Fig. 3–18). These fetuses have minimal survival.

Failure of normal telencephalic cleavage of the forebrain into bilateral cerebral hemispheres results in holoprosencephaly, a midline patterning defect resulting from a disruption of the Sonic hedgehog (SHH) *signaling* pathway. This gives rise to a spectrum of upper facial, eye, nose and ear anomalies, most severe as cyclopia (*synophthalmia*) (Fig. 3–19), and ranging through ethmocephaly, cebocephaly, premaxillary agenesis (median cleft lip/palate), premaxillary dysgenesis (bilateral cleft lip/palate), to the mildest dysmorphic manifestation of a single maxillary central incisor tooth.

**Figure 3–18** Anencephaly, low-set ears and right cleft lip and palate in an aborted fetus. Note absent calvaria.

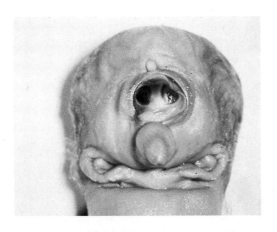

Figure 3–19 An example of combined holoprosencephaly and otocephaly syndromes. Defects include synophthalmia (cyclopia), a median proboscis, synotia and agnathia.

The locations of the eyes, nose and midmaxilla are variably disturbed, producing a wide assortment of dysmorphic syndromes.

Defects of the rhombencephalic organizing center, which is responsible for induction of the viscerofacial skeleton, account for dysmorphology of the middle and lower thirds of the face (otomandibular syndromes). Defective mandibular development may range from agnathia (absent mandible) associated with ventrally placed cervically located ears (*synotia*) (Fig. 3–20) to varying degrees of *micrognathia* and *mandibulofacial dysostosis*. The rare failure of merging of the mandibular prominences results in mandibular midline cleft.

Mild developmental defects of the face are comparatively common. Failure of the facial prominences to merge or fuse results in abnormal developmental clefts. These clefts are due to disruption of the many integrated processes of induction, cell migration, local growth and mesenchymal merging. Unilateral clefting of the upper lip (*cheiloschisis*) is the result of the medial nasal prominence's failure to merge with the maxillary prominences on either side of the midline (Fig. 3–21[a]). Unilateral *cleft*

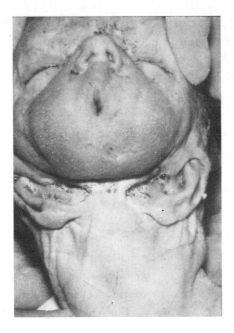

Figure 3–20 A stillborn female 33-week-old fetus with microstomia, aglossia, agnathia and synotia (otocephaly), resulting from aplasia of the first and second pharyngeal arches. (Courtesy of Dr. G. G. Leckie.).

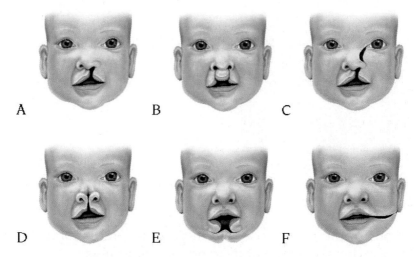

Figure 3–21  Defects of orofacial development: (A) unilateral cleft lip; (B) bilateral cleft lip; (C) oblique facial cleft and cleft lip; (D) median cleft lip and nasal defect; (E) median mandibular cleft; (F) unilateral macrostomia.

*lip*, more usual on the left side, is a relatively common congenital defect (1 in 800 births) that has a strong familial tendency, suggesting a genetic background. Seventy percent of upper lip clefts appear to be isolated phenomena, but frequently are the result of single gene defects in IRF6, MSX1 or FGFR1 (Fig. 3–22). The rare *bilateral cleft lip* results in a wide midline defect of the upper lip, and may produce a protuberant proboscis (Fig. 3–23). The exceedingly rare *median cleft lip* (a true "hare lip") is due to incomplete merging of the two medial nasal prominences and, therefore, in most cases, with deep midline grooving of the nose, leading to various forms of bifid nose.

Merging of the maxillary and mandibular prominences beyond or short of the site for normal mouth size results in a mouth that is too small (*microstomia**) or too wide (*macrostomia†*; see Fig. 3–21[f]). Rarely, the maxillary and mandibular prominences fuse, producing a closed mouth (*astomia*).

An *oblique facial cleft* results from persistence of the groove between the maxillary prominence and the lateral nasal prominence running from the medial canthus of the eye to the ala of the nose. Persistence of the furrow between the two mandibular prominences produces the rare midline *mandibular cleft* (see Fig. 3–21[e]).

Retardation of mandibular development gives rise to micrognathia of varying degrees with accompanying dental malocclusion. Total failure of development of the mandible, *agnathia*, is associated with abnormal ventral placement of the external ears (*synotia*) (see Fig. 3–20).

---

*Such microstomial defects are a common feature of syndromes of congenital anomalies of development; e.g., trisomy 17-18, craniocarpotarsal syndrome (whistling face), the otopalato-digital syndrome, and (occasionally) Turner syndrome.

†Macrostomia occurs in idiopathic hypercalcaemia, mandibulofacial dysostosis (Treacher Collins syndrome), and occasionally in Klinefelter XXY syndrome.

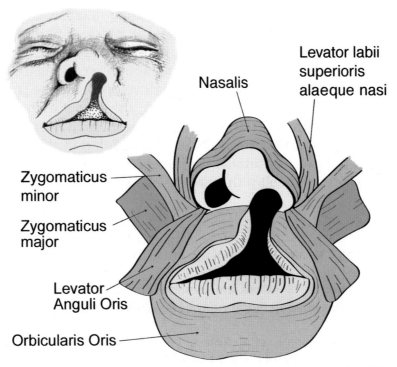

Figure 3–22  Schematic depiction of muscle orientation in unilateral cleft lip.

*Craniofacial developmental cysts*, although strictly speaking not a defect of orofacial development, originate in the complicated embryonic processes of the craniofacial complex. Developmental cysts arise along the lines of facial and palatal clefts, and their lining epithelia appear to be derived from residues, or "rests," of the covering epithelia of the embryonic prominences that merge to form the face. Where such epithelial residues become trapped in the subjacent mesenchyme during merging, or ectopic sequestration of skin or mucosa occurs beneath the surface, there is a potential for cyst formation. In most instances, the subsurface epithelium degenerates, probably by programmed cell death. Persisting epithelial

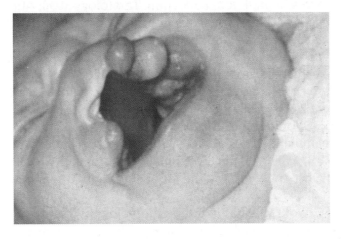

Figure 3–23  Bilateral cleft lip and palate, with consequent projection of the median globular process, in a newborn infant.

rests may be stimulated to proliferate; after their necrosis, followed by a period of dormancy; in postnatal life they can produce fluid-filled cysts. The nature and cause of stimuli that give rise to these developmental cysts are unknown. These cysts tend to be named according to the site in which they develop. Hence, *nasolabial* cysts develop where the lateral nasal prominence and maxillary prominence meet; *globulomaxillary cysts* develop more deeply along the line of merging of the median nasal and maxillary prominences and *median mandibular cysts* develop in the midline site of merging of the two mandibular prominences. Cysts of the pharyngeal arches and palate are dealt with respectively in Chapters 5 and 10.

## SELECTED BIBLIOGRAPHY

Apesos J, Anigian GM. Median cleft of the lip: its significance and surgical repair. Cleft Palate-Craniofac J 1993; 30:94–96.

Balmer CW, LaMantia AS. Noses and neurons: induction, morphogenesis and neuronal differentiation in the peripheral olfactory pathway. Dev Dyn 2005; 234:464–481.

Barni T, Fantoni G, Gloria L, et al. Role of endothelin in the human craniofacial morphogenesis. J Craniofac Genet Dev Biol 1998; 18:183–194.

Ben Ami M, Weiner E, Perlitz Y, Shalev E. Ultrasound evaluation of the width of the fetal nose. Prenatal Diag 1998; 18:1010–1013.

Bhattacharya S, Bronner-Fraser M. Competence, specification and commitment to an olfactory placode fate. Development 2008; 135: 4165–4177.

Boehm N, Gasser B. Sensory receptor-like cells in the human foetal vomeronasal organ. Neuroreport 1993; 4:867–870.

Diewert VM, Lozanoff S. Growth and morphogenesis of the human embryonic midface during primary palate formation analyzed in frontal sections. J Craniofac Genet Dev Biol 1993; 13:162–183.

Diewert VM, Lozanoff S, Choy V. Computer reconstructions of human embryonic craniofacial morphology showing changes in relations between the face and brain during primary palate formation. J Craniofac Genet Dev Biol 1993; 13:193–201.

Diewert VM, Shiota K. Morphological observations in normal primary palate and cleft lip embryos in the Kyoto collection. Teratology 1990; 41:663–677.

Diewert VM, Wang KY. Recent advances in primary palate and midface morphogenesis research. Crit Rev Oral Biol Med 1992; 4:111–130.

Garrosa M, Gayoso MJ, Esteban FJ. Prenatal development of the mammalian vomeronasal organ. Microsc Res Tech 1998; 41:456–470.

Helms JA, Cordero D, Tapadia MD. New insights into craniofacial morphogenesis. Development 2005; 132:851–861.

Hu D, Marcucio RS. Unique organization of the frontonasal ectodermal zone in birds and mammals. Dev Biol 2009; 3251: 200–210.

Hu D, Marcucio RS. A SHH-responsive signaling center in the forebrain regulates craniofacial morphogenesis via the facial ectoderm. Development 2009; 136: 107–116.

Kehrli P, Maillot C, Wolff MJ. Anatomy and embryology of the trigeminal nerve and its branches in the parasellar area. Neurol Res 1997; 19:57–65.

Kjaer I, Fischer-Hansen B. Human fetal pituitary gland in holoprosencephaly and anencephaly. J Craniofac Genet Dev Biol 1995; 15:222–229.

Kjaer I, Fischer-Hansen B. The human vomeronasal organ: prenatal developmental stages and distribution of luteinizing hormone-releasing hormone. Eur J Oral Sci 1996; 104:34–40.

Kosaka K, Hama K, Eto K. Light and electron microscopy study of fusion of facial prominences. A distinctive type of superficial cells at the contact sites. Anat Embryol 1985; 173:187–201.

Lekkas C, Latief BS, Corputty JE. Median cleft of the lower lip associated with lip pits and cleft of the lip and palate. Cleft Palate-Craniofac J 1998; 35:269–271.

Li W-Y, Dudas, Kaartinen V. Signaling through Tgf-B type 1 receptor Alk5 is required for upper lip fusion. Mech Dev 2008; 125: 874–882.

McCabe KL, Bronner-Fraser M. Molecular and tissue interactions governing induction of cranial ectodermal placodes. Dev Biol 2009; 332:189–195

McCabe KL, Bronner-Fraser M. PDGF signaling is critical for trigeminal placode formation. Dev Biol 2008; 319:534.

Mooney MP, Siegel MI, Kimes KR, Todhunter J. Premaxillary development in normal and cleft lip and palate human fetuses using three-dimensional computer reconstruction. Cleft Palate-Craniofac J 1991; 28:49–53, discussion 54.

Mukhopadhyay P, Greene RM, Pisano MM. Expression profiling of transforming growth factor beta superfamily genes in developing orofacial tissue. Birth Defects Res A Clin Mol Teratol 2006; 76:528–543.

Namnoum JD, Hisley KC, Graepel S, et al. Three-dimensional reconstruction of the human fetal philtrum. Ann Plast Surg; 1997; 38:202–208.

Oostrom CA, Vermeij-Keers C, Gilbert PM, van der Meulen JC. Median cleft of the lower lip and mandible: case reports, a new embryologic hypothesis, and subdivision. Plast Reconst Surg 1996; 97:313–320.

Pretorius DH, Nelson TR. Fetal face visualization using three-dimensional ultrasonography. J Ultrasound Med 1995; 14:349–356.

Sataloff RT, Sieber JC. Phylogeny and embryology of the facial nerve and related structures. Ear, Nose, Throat J 2003; 9:704–724; 10:764–779.

Schwartz Q, Waimey KE, Golding M, et al. Plexin A3 and plexin A4 convey semaphorin signals during facial nerve development. Dev Biol 2008; 324:1–9.

Sherwood RJ, McLachlan JC, Aiton JF, Scarborough J. The vomeronasal organ in the human embryo, studied by means of three-dimensional computer reconstruction. J Anat 1999; 195:413–418.

Siegel MI, Mooney MP, Kimes KR, Todhunter J. Developmental correlates of midfacial components in a normal and cleft lip and palate human fetal sample. Cleft Palate-Craniofac J 1991; 28:408–412.

Smith TD, Siegel MI, Mooney MP, et al. Vomeronasal organ growth and development in normal and cleft lip and palate human fetuses. Cleft Palate-Craniofac J 1996; 33:385–394.

Smith TD, Siegel MI, Mooney MP, et al. Prenatal growth of the human vomeronasal organ. Anat Rec 1997; 248:447–455.

Song, L, Li Y, Wang K, et al. Lrp6-mediated canonical Wnt signaling is required for lip formation and fusion. Development 2009; 136:3161-3171.

Sperber GH, Sperber SM. Embryology of Ororfacial Clefting. In Losee J, Kirschner R (eds). Comprehensive cleft care. New York: McGraw-Hill, 2009:3–20.

Sulik KK. Dr. Beverly R. Rollnick memorial lecture. Normal and abnormal craniofacial embryogenesis. Birth Defects: Original Article Series 1990; 26:1–18.

Sulik KK, Schoenwolf GC. Highlights of craniofacial morphogenesis in mammalian embryos, as revealed by scanning electron microscopy. Scan Elect Micro
1985; (Pt 4):1735–1752.

Thorogood P, Ferretti P. Heads and tales: recent advances in craniofacial development. Br Dent J 1992; 173:301–306.

Szabo-Rogers HL, Geetha-Lognathan P, Whiting CJ, et al. Novel skeletogenic patterning roles for the olfactory pit. Development 2009; 136:219–229.

Thomason HA, Dixon MJ, Dixon J. Facial clefting in Tp63 deficient mice results
from altered Bmp4, Fgf8 and Shh *signaling*. Dev Biol 2008; 321: 273–282.

Varga ZM, Wegner J, Westerfield M. Anterior movement of ventral diencephalic
precursors separates the primordial eye field in the neural plate and requires cyclops. Development 1999; 126:5533–5546.

Williams A, Pizzuto M, Brodsky L, Perry R. Supernumerary nostril: a rare congenital deformity. Int J Pediatr Otorhinolaryngol1998; 44:161–167.

Yonemoto H, Yoshida K, Kinoshita K, Takeuchi H. Embryological evaluation of
surface features of human embryos and early fetuses by 3-D ultrasound. J
Obstet Gynecol Res 2002; 28:211–216.

# 4 Pharyngeal Arches

*Starting from a single cell, I passed one period of my life with gill slits inherited from my fishy ancestry, then for a few weeks sported a tail and was hard to distinguish from an unborn tree shrew....Why think of viruses or pre-Cambrian organisms, when inside this delicate membrane of my skin, this outline of an individual, I carry the whole history of life.*

Jaquetta Hawkes (1953)

## NORMAL DEVELOPMENT

During the late somite period (fourth week postconception [pc]), the mesoderm lateral plate of the ventral foregut region becomes segmented to form a series of five distinct bilateral mesenchyme swellings, the *pharyngeal (branchial) arches*. Ventrally migrating neural crest cells interact with lateral extensions of the pharyngeal endoderm, surround the six aortic arch arteries, and initiate pharyngeal arch development. The initial mesodermal core of each arch is augmented by neural crest tissue that surrounds the mesodermal core. The mesoderm will give rise to muscle myoblasts, while the neural crest cells give rise to skeletal and connective tissues.

The *pharyngeal arches* are separated by *pharyngeal grooves* on the external aspect of the embryo, which correspond internally with five outpouchings of the elongated pharynx of the foregut, known as the five *pharyngeal pouches* (Figs. 4–1 and 4–2). Although derivatives of five or even six arches are

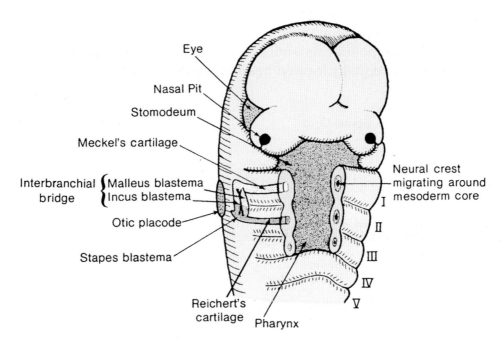

Figure 4–1 Schematic diagram of embryo with sectioned pharyngeal arches.

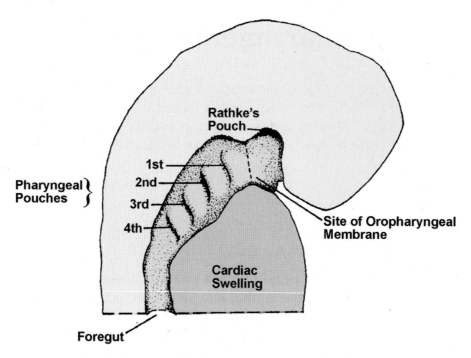

**Figure 4–2** Internal representation of the pharyngeal pouches (numbered) in a somite-period embryo.

described, only three (and exceptionally, four) arches appear externally. Caudal to the third arch there is a depression, the cervical sinus.

The pharyngeal arches decrease in size from cranial to caudal, each pair merging midventrally to form "collars" in the cervical region. Each of the five pairs of arches contains a basic set of structures (Figs. 4–3 to 4–7; Table 4–1):

1. A central *cartilage* rod that forms the skeleton of the arch
2. A muscular component
3. A vascular component, an *aortic arch artery* that runs around the pharynx from the ventrally located heart to the dorsal aorta

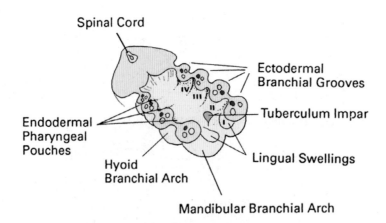

**Figure 4–3** Schema of early pharyngeal arch development. Pharyngeal arches numbered in Roman numerals. (After Waterman and Meller.)

## TABLE 4–1: Derivatives of the Pharyngeal System

| Pharyngeal Arch | Ectodermal Groove | Endodermal Pouch | Skeleton | Viscera | Artery | Muscles | Motor Nerve | Sensory Nerve |
|---|---|---|---|---|---|---|---|---|
| 1st Mandibular | External acoustic meatus: ear hillocks; pinna | Auditory tube; middle ear tympanum | Meckel's cartilage: malleus; incus; (mandible template) | Body of tongue | External carotid artery Maxillary artery | Masticatory, tensor tympani, tensor palatini, mylohyoid, anterior digastric | Mand. Division of V Trigeminal | V Lingual nerve |
| 2nd Hyoid | Disappears | Tonsillar fossa | Stapes, styloid process; superior hyoid body | Midtongue: Thyroid gland anlage; tonsil | Stapedial artery (disappears) | Facial, stapedius, hyoid, posterior digastric | VII Facial | VII Chorda tympani |
| 3rd | Disappears | Inferior parathyroid 3; thymus | Inferior hyoid body, great cornu hyoid | Root of tongue: fauces; epiglottis; thymus Inferior parathyroid 3; carotid body | Internal carotid artery | Stylopharyngeus | IX Glossopharyngeal | IX Glossopharyngeal |
| 4th | Disappears | Superior parathyroid 4 | Thyroid & laryngeal cartilages | Pharynx: epiglottis Superior parathyroid 4 Para-aortic bodies | Aorta (L) Subclavian (R) | Pharyngeal constrictors; Levator palatini Palatoglossus Palatopharyngeus | X Superior laryngeal nerve Vagus | X Auricular nerve to external acoustic meatus |
| 6th | Disappears | Telopharyngeal (ultimopharyngeal) body (cyst) Calcitonin "C" cells | Cricoid, arytenoid, corniculate cartilages | Larynx | Pulmonary arteries; Ductus arteriosus | Cricothyroid Laryngeal muscles Pharyngeal constrictors | X inferior laryngeal nerve Vagus | X Vagus |
| Postpharyngeal region | | | Tracheal cartilages | | | Trapezius; Sternomastoid | XI Spinal accessory | |
| Somites 4 Occipital somites | | | Sclerotomes Basiocipital bone | | | Myotomic muscle: Intrinsic tongue muscles: Extrinsic tongue muscles: Styloglossus Hyoglossus, Genioglossus muscles | XII Hypoglossal | |
| Prechordal somites | | | Nasal septum Nasal capsule | | | Extrinsic ocular muscles | III. Oculomotor IV Trochlear VI Abducens | |
| Upper cervical somites | | | Cervical vertebrae | | | Geniohyoid; Infrahyoid muscles | Spinal nerves C1, C2 | |

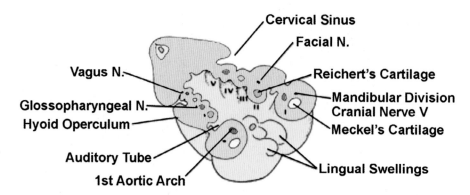

Figure 4–4  Schema of lateral pharyngeal arch development. Pharyngeal arches numbered in Roman numerals. (After Waterman and Meller.)

4. A nervous element, consisting of sensory and special visceral motor fibers of one or more cranial nerves supplying the mucosa and muscle arising from that arch

   The cartilage rods, which are differentiated from neural crest tissue that has been organized by pharyngeal endoderm, are variously adapted to bony, cartilaginous or ligamentous structures, or in some cases may disappear in later development. The muscle components are of somitomeric and somitic origin and give rise to special visceral muscles composed of striated muscle fibers. Nerve fibers of specific cranial nerves enter the mesoderm of the pharyngeal arches, initiating muscle development in the mesoderm. The muscles arising from the mesodermal cores migrate from their sites of origin and adapt to the pharyngeal arch derivatives. The original nerve supply to these muscles is maintained during migration,

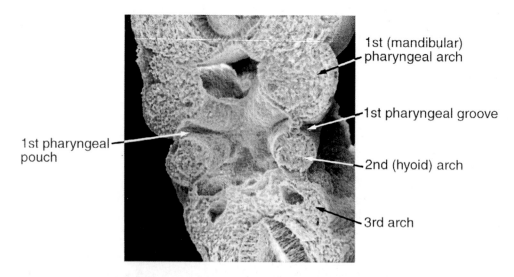

Figure 4–5  Scanning electron micrograph of a coronally sectioned pharyngeal region of a mouse embryo corresponding to approximately 28-day-old human embryo. (Courtesy of Dr. K. K. Sulik, University of North Carolina.)

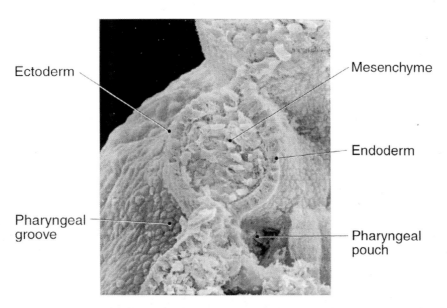

Ectoderm

Mesenchyme

Endoderm

Pharyngeal
groove

Pharyngeal
pouch

**Figure 4–6** Magnified scanning electron micrograph of coronally sectioned pharyngeal arch of a 9-day-old mouse embryo, corresponding to a 28-day-old human embryo. Details of cellular components. (Courtesy of Dr. K. K. Sulik, University of North Carolina.)

accounting for the devious routes of many cranial nerves in adult anatomy. The vascular system originates from lateral plate mesoderm and neural crest tissue angioblasts. The arteries are modified from their symmetrical primitive embryonic pattern into the asymmetrical form of the adult derivatives of the fourth and sixth arch arteries, through the signaling of TBX1 on GBX2 expression.

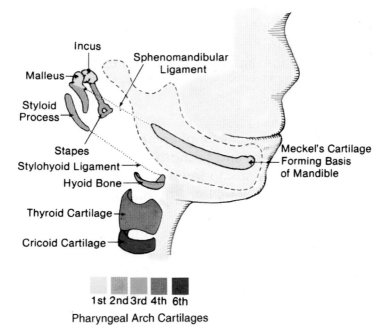

Incus

Sphenomandibular
Ligament

Malleus

Styloid
Process

Stapes

Stylohyoid Ligament

Hyoid Bone

Thyroid Cartilage

Cricoid Cartilage

Meckel's Cartilage
Forming Basis
of Mandible

1st 2nd 3rd 4th 6th
Pharyngeal Arch Cartilages

**Figure 4–7** Derivatives of the pharyngeal arch cartilages. Note that the malleus, incus and hyoid bones are each derived from two pharyngeal arches.

### First Pharyngeal Arch

*Genetics*

The first pharyngeal arch is crucial to the development of the face, and identification of the genes expressed during its development is significant to understanding normal and abnormal facial formation. The paired arches appear during the fourth and fifth weeks postconception, when certain gene expression patterns are manifested. Three hundred and fifteen genes have been shown to be expressed in the first pharyngeal arch during this time. Some 97 genes are differentially expressed between the fourth and fifth weeks. These genes account for different molecular functions and different biological processes. These include signal transductions, transcription regulations, binding and catalytic activities. The genes identified include MSX1, MSX2, DLX2, PAX8, PAX 9, BMP4, BMP5, BMP7, FGFR11 and TGFB1. The first and second (mandibular and hyoid) arches are devoid of any HOX gene activity. HOXA2 and HOXA3 act synergistically to pattern the third and fourth arches. The BMP4 and MSX2 gene expressions in migrating neural crest cells are instigators of apoptosis (cell death) that create the grooves separating the arches. It is the lack of HOX gene expression that determines the identity of the cranial neural crest that migrates to form the first pharyngeal arch, and subsquently the mandible and its associated structures.

It is at this early stage that the rostrocaudal polarization of the first arch axes is established. Reciprocal epithelial and mesenchymal *signaling* events induce transcription factors that include GSC, LHX and DLX genes that function to differentiate the oral versus aboral rostrocaudal surfaces, as well as the proximal and distal axes that presage the odontogenic homeodomain code and pattern the subsequent skeletal elements. The regionalization of the first arch occurs through antagonistic *signaling* mechanisms of BMP and FGF morphogenetic cues. Downstream transcription factors that include MSX1/2, BARX1 and DLX genes expression patterns presage the incisivization and molarization zones that will ultimately pattern the teeth.

*Morphogenesis*

The first, or mandibular, pair of pharyngeal arches are the precursors of the jaws, both maxillary and mandibular, and appropriately bound the lateral aspects of the stomodeum, which at this stage is merely a depression in the early facial region. The maxilla is derived from a small *maxillary* prominence extending cranioventrally from the much larger *mandibular* prominence derived from the first arch. The cartilage skeleton of the first arch, known as *Meckel's cartilage* (see Fig. 4–1), which arises at the 41st to 45th days pc, provides a template for subsequent development of the mandible, but most of its cartilage substance disappears in the formed mandible. The mental ossicle is the only portion of the mandible derived from Meckel's cartilage by endochondral ossification.

Persisting portions of Meckel's cartilage form the basis of major portions of two ear ossicles, the head and neck of the *malleus\** and the body and short crus of the *incus*[†] and two ligaments—the *anterior ligament of the malleus* and the *sphenomandibular ligament*. The musculature of the mandibular arch, originating from cranial somitomere 4, subdivides and migrates to form the *muscles of mastication*, the *mylohyoid muscle*, the *anterior belly of the digastric*, the *tensor tympani*,[‡] and the *tensor veli palatini muscles*, all of which are innervated by the nerve of the first arch, namely, the *mandibular division of the fifth cranial or trigeminal* nerve (see Fig. 4–4). The sensory component of this nerve innervates the mandible and its covering mucosa and gingiva, the mandibular teeth, the mucosa of the anterior two-thirds of the tongue, the floor of the mouth, and the skin of the lower third of the face. The first arch artery contributes, in part, to the maxillary artery and part of the *external carotid artery*.

## Second Pharyngeal Arch

The cartilage of the second or *hyoid arch* (Reichert's cartilage), appearing at the 45th to 48th days pc, is the basis of the greater part (head, neck and crura) of the third ear ossicle, the *stapes,\*\** and contributes to the malleus and incus,[†] the *styloid process* of the temporal bone, the *stylohyoid ligament*, and the *lesser horn* and cranial part of the *body of the hyoid bone* (Fig 4–7). Reichert's cartilage attaches to the basicranial otic capsule, where it is grooved by the facial nerve; it provides the remaining cartilaginous circumference to the labyrinthine and tympanic segments of the facial canal.

The muscles of the hyoid arch originating from cranial somitomere 6 subdivide and migrate extensively to form the *stapedius*, the *stylohyoid*, the posterior *belly of the digastric*, and the *mimetic muscles* of the face, all of which are innervated by the *VIIth cranial* or *facial nerve*, serving the second arch. The paths of migration of these muscles are traced out in the adult by the distribution of branches of the facial nerve. The special sensory component of this nerve for taste, known as the *chorda tympani nerve*, invades the first arch as a pretrematic nerve and thus comes to supply the mucosa of the anterior two-thirds of the tongue. The artery of this arch forms the

---

\*The anterior process of the malleus forms independently in membrane bone (os goniale).

[†]The incus arises from the separated dorsal end of Meckel's cartilage that corresponds to the pterygoquadrate cartilage of inframammalian vertebrates.

[‡]The tensor tympani is the evolutionary remnant of a reptilian jaw muscle attached to the remnant of the reptilian jaw, i.e., Meckel's cartilage. The stapes is the first ear ossicle to appear. The footplate (base) of the stapes is derived in part from the lateral wall of the otic capsule.

\*\*The first ossicle to appear. The footplate (base) of the stapes is derived in part from the lateral wall of the otic capsule.

[†]The manubrium of the malleus and the long crus of the incus are derived from the interpharyngeal bridge, which arises from the second arch.

*stapedial artery*, which disappears during the fetal period, leaving the foramen in the stapes.

The stapedial artery, derived from the second aortic arch, is significant in the development of the stapes ear ossicle. The stapedial blastema grows around the stapedial artery, forming a ring around the centrally placed artery. The midportion of the stapedial artery involutes, leaving the foramen in the stapes. Stapedial arterial branches persist to become part of the internal carotid artery proximally and the external carotid artery distally.

### Third Pharyngeal Arch

The cartilage of this small arch produces the greater horn and caudal part of the body of the hyoid bone. The remainder of the cartilage disappears. The mesoderm originating from cranial somitomere 7 forms the *stylopharyngeus muscle*, innervated by the IXth (*glossopharyngeal*) nerve supplying the arch. The mucosa of the posterior third of the tongue is derived from this arch, which accounts for its sensory innervation by the glossopharyngeal nerve.

The artery of this arch contributes to the *common carotid* and part of the *internal carotid arteries*. Neural crest tissue in the third arch forms the *carotid body* which first appears as a mesenchymal condensation around the third aortic arch artery. This chemoreceptor body thus derives its nerve supply from the glossopharyngeal nerve.

### Fourth Pharyngeal Arch

The cartilage of this arch probably forms the thyroid cartilage. The arch muscles originating from occipital somites 1 and 2 develop into the *cricothyroid* and *constrictors of the pharynx*, the *palatopharyngeus*, *levator veli palatini* and *uvular muscles* of the soft palate and the *palatoglossus muscle* of the tongue. The nerve of the fourth arch is the superior laryngeal branch of the vagus (Xth cranial) nerve, which innervates these muscles.

The fourth-arch artery of the left side forms the arch of the *aorta*; that of the right side contributes to the *right subclavian* and *brachiocephalic* arteries. The *para-aortic bodies* of chromaffin cells that secrete noradrenalin arise from the ectomesenchyme of the fourth and sixth pharyngeal arches.

### Fifth Pharyngeal Arch

The fifth arch, a transitory structure, disappears almost as soon as it forms, and bequeaths no permanent structural elements.

### Sixth Pharyngeal Arch

The cartilage of this arch probably forms the *cricoid* and *arytenoid* cartilages of the larynx. The mesoderm originating from occipital somites 1 and 2 forms the intrinsic muscles of the larynx, which are supplied by the nerve of the arch, the *recurrent laryngeal branch of the Xth cranial* or *vagus nerve*. Because the nerve of this arch passes caudal to the fourth-arch artery in its recurring path from the brain to the muscles it supplies, the differing fate of the left and right fourth-arch arteries, when migrating caudally into the thorax, accounts for the different recurrent paths of the left and right laryngeal nerves. The right recurrent laryngeal nerve recurves around the right subclavian artery (derived from the fourth arch artery); the left laryngeal nerve recurves around the aorta (derived from the fourth-arch artery), due to the dorsal part of the sixth aortic artery and all of the fifth aortic artery having disappeared.

The sixth-arch arteries develop in part into the *pulmonary arteries*, the remainder disappearing on the right side, and forming on the left side the temporary *ductus arteriosus* of the fetal circulation that becomes the *ligamentum arteriosum*.

### Postpharyngeal Region

Controversy surrounds the embryological origin of the tracheal cartilages and the sternomastoid and trapezius muscles. On the basis of their nerve supply (spinal accessory cranial nerve XI), it appears that the latter two muscles are of mixed somitic and pharyngeal arch origin.

### Hyoid Bone

This composite* endochondral bone derived from the cartilages of the second (hyoid) and third pharyngeal arches reflects its double origin in its six centers of ossification, two for the body and one for each lesser and each greater horn. Ossification centers for the greater horns appear at the end of fetal life (38 weeks), those for the body at about birth, and those for the lesser horns, about 2 years postnatally. The lesser horns may not fuse with the body, but attach by fibrous tissue to the major horns that in turn articulate with the body through a diarthrodial synovial joint. The double origin of the hyoid body, from the second and third arches, is very occasionally reflected as a bone split into upper and lower portions. Fusion of all these elements into a single bone occurs in early childhood.

## ANOMALIES OF DEVELOPMENT

Deficient development of the pharyngeal arches results in syndromes that are identified according to the arch(es) involved. Thus, there are first-, second-, and so on, arch syndromes, each subsequent one rarer, reflecting deficiencies of all or some of the derivatives of that arch. Examples of severe first-arch anomalies are agnathia, synotia and microstomia.

---

*The hyoid bone consists of several independent elements that, although fused together in humans, remain separate in many animals as the "hyoid apparatus." Of these separate elements, the tympanohyal and stylohyal form the styloid process in humans. The ceratohyal forms the lesser horn and upper part of the body of the human hyoid. The epihyal does not form a bone in humans, but is incorporated into the stylohyoid ligament. The thyrohyal that develops from the third arch forms the greater horn and lower part of the body of the human hyoid bone.

**TABLE 4–2: Possible Anomalies of the Pharyngeal System**

| Arch | Ectodermal Groove | Endodermal Pouch | Skeleton | Artery | Muscles | Nerve |
|---|---|---|---|---|---|---|
| 1st Mandibular | Aplasia, atresia, stenosis, duplication of external acoustic meatus | Diverticulum of auditory tube; aplasia, atresia, stenosis of tube | Aplasia/dysplasia of malleus, incus, mandible | Hypoplastic/absent external carotid & maxillary arteries | Deficient masticatory/facial muscles | Absent mandibular nerve |
| 2nd Hyoid | Cervical (pharyngeal) cleft, sinus, cyst, fistula | Tonsillar sinus, pharyngeal fistula | Aplasia/dysplasia of stapes, styloid process | Persistent stapedial artery | Deficient facial, stapedial muscles | Deficient facial, chorda tympani nerve |
| 3rd | Cervical cleft, sinus, cyst, fistula | Cervical thymus: thymic cyst; aplasia parathyroid 3; aplasia thymus (DiGeorge anomaly) | Defective hyoid bone | Hypoplastic/absent internal carotid artery | | Deficient glossopharyngeal nerve |
| 4th | Cervical cleft, sinus, cyst, fistula | Fistula/sinus from pyriform sinus; aplasia parathyroid 4 | Congenital laryngeal stenosis, cleft, atresia (Fraser syndrome) | Double aortae: aortic interruption; right aorta | Deficient faucial muscles | Deficient vagus nerve |
| 6th | | Aplasia of calcitonin "C" cells (DiGeorge anomaly) | | Aorticopulmonary septation anomalies (DiGeorge anomaly) | | |

Less severe anomalies are mandibulofacial dysostosis (Treacher Collins syndrome) and micrognathia combined with cleft palate (Pierre Robin syndrome). The pathogenesis of Treacher Collins syndrome arises from a haploinsufficiency of the TREACLE protein encoded by the TCOF1 gene that causes apoptosis of the cranial neural crest resulting in a deficiency of the midface and jaws. External ear deficiencies (anotia, microtia), auricular tags and persistent pharyngeal clefts or cysts (auricular sinuses) are among the commoner examples of pharyngeal arch anomalies.

Anomalies of the second and subsequent arches involve the hyoid (laryngeal) apparatus and are very rare. The rare autosomal dominant branchio-oculofacial syndrome (BOFS [OMIM 113620]) involving the first and second pharyngeal arches has been ascribed to a mutation of TFAP2α on chromosome 6p24.3.

Mineralization of the stylohyoid ligament elongates the styloid process that may cause craniocervical pain, dysphagia, odynophagia and foreign body discomfort of the pharynx (Eagle syndrome). The condition may be inherited by an autosomal dominant gene on chromosome 6p (Table 4–2).

## SELECTED BIBLIOGRAPHY

Bamforth JS, Machin GA. Severe hemifacial microsomia and absent right pharyngeal arch artery derivatives in a 19-week-old fetus. Birth Defects: Orig Art Series 1996; 30:227–245.

Cai J, Ash D, Kotch LE, et al. Gene expression in pharyngeal arch 1 during human embryonic development. Hum Mol Genet 2005; 14:903–912.

Calmont A, Ivins S, Van Bueren KL et al. Tbx1 controls cardiac neural crest migration during arch artery development by regulating Gbx2 expression in the pharyngeal ectoderm. Development 2009; 136:3173–3183.

Clementi M, Mammi I, Tenconi R. Family with branchial arch anomalies, hearing loss, ear and commissural lip pits, and rib anomalies. A new autosomal recessive condition: branchio-oto-costal syndrome. Am J Med Genet 1997; 68:91–93.

Crump JG, Swartz ME, Kimmel CB. An integrin-dependent role of pouch endoderm in hyoid cartilage development. PLos Biol 2004; 2:e244.

Ferguson CA, Tucker AS, Sharpe PT. Temporospatial cell interactions regulating mandibular and maxillary arch patterning. Development 2000; 127:403–412.

Gavalas A, Studer M, Lumsden A, et al. Hoxa1 and Hoxb1 synergize in patterning the hindbrain, cranial nerves and second pharyngeal arch; Development 1998; 125:1123–1136.

Graham A, Smith A. Patterning the pharyngeal arches. Bioessays 2001; 23:54–61

Hunt P, Whiting J, Muchamore I, et al. Homeobox genes and models for patterning the hindbrain and branchial arches. Development Suppl 1991; 1:187–196.

Jacobsson C, Granstrom G. Clinical appearance of spontaneous and induced first and second branchial arch syndromes. Scand J Plastic Reconst Surg Hand Surg 1997; 31:125–136.

Kruchinskii GV. Classification of the syndromes of branchial arches 1 and 2. Acta Chir Plast 1990; 32:178–190.

Merida-Velasco JA, Sanchez-Montesinos I, Espin-Ferra J, et al. Ectodermal ablation of the third branchial arch in chick embryos and the morphogenesis of the parathyroid III gland. J Craniofac Genet Dev Biol 1999; 9:33–40.

Milunsky JM, Maher TA, Zhao G, et al. TFAP2A Mutations result in branchio-oculofacial syndrome. Am J Hum Genet 2008; 82:1171–1177.

Minoux M, Antonarakis GS, Kmita M, et al. Rostral and caudal pharyngeal arches share a common neural crest ground pattern. Development 2009; 136:637–645.

Passos-Bueno MR, Ornelas CC, Fanganiello RD. Syndromes of the first and second pharyngeal arches: a review. Amer J Med Genet 2009; 149A:1853–1859.

Pearl WR. Single arterial trunk arising from the aortic arch. Evidence that the fifth branchial arch can persist as the definitive aortic arch. Pediatr Radiol 1991; 21:518–520.

Pretterklieber ML, Krammer EB. Sphenoidal artery, ramus orbitalis persistens and pterygospinosus muscle–a unique cooccurrence of first branchial arch anomalies in man. Acta Anat 1996; 155:136–144.

Qiu M, Bulfone A, Ghattas I, et al. Role of the Dlx homeobox genes in proximo-distal patterning of the branchial arches: mutations of Dlx-1. Dev Biol 1997; 185(2):165–164.

Stratakis CA, Lin JP, Rennert OM. Description of a large kindred with autosomal dominant inheritance of branchial arch anomalies, hearing loss, and ear pits, and exclusion of the branchio-oto-renal (BOR) syndrome gene locus (chromosome 8q13.3). Am J Med Genet 1998; 79:209–214.

Tucker AS, Yamada G, Grigoriou M, et al. Fgf-8 determines rostral-caudal polarity in the first branchial arch. Development 1999; 126:51–61.

Veitch E, Begbie J, Schilling TF, et al. Pharyngeal arch patterning in the absence of neural crest. Curr Biol 1999; 9:1481–14844.

Vieille-Grosjean I, Hunt P, Gulisano M, et al. Branchial HOX gene expression and human craniofacial development. Dev Biol 1997; 183:49–60.

Whitworth IH, Suvarna SK, Wight RG, Walsh-Waring GP. Fourth branchial arch anomaly: a rare incidental finding in an adult. J Laryng Otol 1993; 107:238–239.

# 5 Pharyngeal Pouches and Pharyngeal Grooves

## NORMAL DEVELOPMENT

The primitive pharynx forms in the late embryonic period as a dilation of the cranial end of the foregut, lying between the developing heart ventrally and the developing chondrocranium rostrodorsally. The early pharynx is large relative to the rest of the gut, is flattened ventrodorsally, and gives rise to diverse structures from its floor and side walls. The lateral aspects of the comparatively elongated primitive pharynx project a series of pouches between the pharyngeal arches, *pharyngeal pouches*, that sequentially decrease in size craniocaudally (Fig. 5–1). Intervening between the pharyngeal arches externally are the pharyngeal grooves (ectodermal clefts) (see Figs 4–1, 4–3 and 4–4). The lining of the pharyngeal grooves is the surface ectoderm, and of the internal pharyngeal pouches is foregut endoderm. Each ectodermal pharyngeal groove corresponds with each endodermal pharyngeal pouch, with a layer of mesodermal mesenchyme intervening between the outer and inner primary germ layers. The endodermal lining of the primitive pharynx develops gradually from a polyhedral cuboidal embryonic epithelium into a respiratory mucous membrane, characterized by a ciliated columnar epithelium and goblet cells.

The first pharyngeal groove persists and while its ventral end is obliterated, its dorsal end deepens to form the *external acoustic meatus*. The

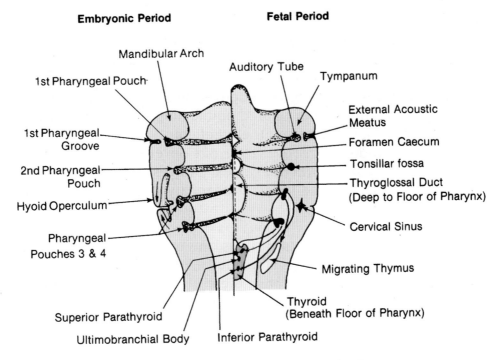

**Figure 5–1** Schema of pharyngeal pouch and pharyngeal groove development.

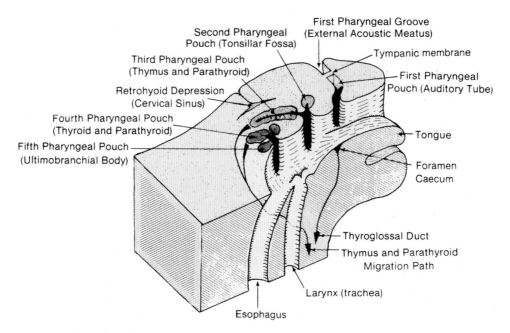

**Figure 5–2** Schematic depiction of pharyngeal pouch derivatives and their migration paths.

ectomesoendodermal membrane in the depth of the groove, separating it from the first pharyngeal pouch, persists as the *tympanic membrane* (Fig. 5–2). Subsequent external and middle ear development is described in Chapter 18.

The second, third and fourth pharyngeal grooves become obliterated by the caudal overgrowth of the second pharyngeal arch (*hyoid operculum*), which provides a smooth contour to the neck. At the end of the fifth week postconception (pc), the third and fourth pharyngeal arches are collectively sunk into a retrohyoid depression, the *cervical sinus* (Fig. 5–3). Failure to obliterate completely these pharyngeal grooves results in a *pharyngeal fistula* leading from the pharynx to the outside, or a *pharyngeal (cervical) sinus* or *cyst*, forming a closed sac (Fig 5–4). Most pharyngeal cysts and fistulae originate from the second groove. If derived from the pharyngeal pouch, they are lined by columnar or ciliated epithelium; if from the pharyngeal groove, they are lined by squamous epithelium.

Thickened ectodermal *epibranchial placodes* develop at the dorsal ends of the first, second and fourth pharyngeal grooves. These placodes contribute to the ganglia of the facial, glossopharyngeal and vagus nerves (see p. 52).

The five pairs of pharyngeal pouches on the sides of the pharyngeal foregut form dorsal and ventral pockets, the endodermal epithelium of which differentiates into various structures. Elongation of the third, fourth and fifth pharyngeal pouches during the sixth and seventh weeks pc increasingly dissociates the pouches from the pharynx, allowing their derivatives to form in the lower anterior neck region (Fig. 5–4; see Figs. 5–1, 5–2, and Table 4–1).

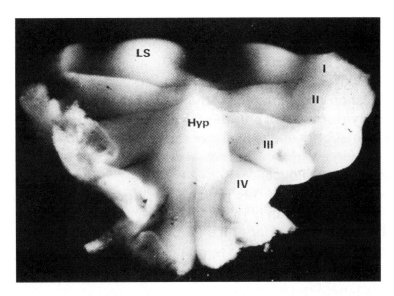

Figure 5–3  The ventral pharyngeal wall of a 32-day-old embryo. The pharyngeal arches are marked I–IV. The lingual swellings (LS) are rising from the first pharyngeal arches. The hypopharyngeal eminence (Hyp) forms a prominent central elevation from which the epiglottis will arise. (× 41, reduced to 75% on reproduction.) (Courtesy of Professor H. Nishimura.)

## First Pharyngeal Pouch

The ventral portion of this pouch is obliterated by the developing tongue. The dorsal diverticulum deepens laterally as the *tubotympanic recess* to form the *auditory tube,* widening at its end into the *tympanum,* or middle ear cavity, separated from the first pharyngeal groove by the tympanic membrane (see Figs. 5–1 and 5–2). The tympanum becomes occupied by the dorsal ends of the cartilages of the first and second pharyngeal arches that develop into the ear ossicles. The tympanum maintains contact with the pharynx via the auditory tube throughout life.

The proximal portion of the expanding and elongating auditory tube becomes lined with respiratory mucous membrane, and fibrous tissue and

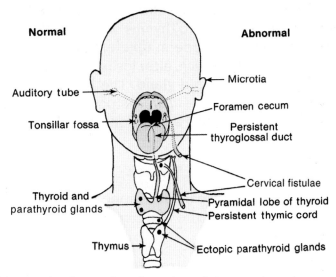

Figure 5–4  Adult normal and anomalous derivatives of pharyngeal pouches.

cartilage form in its walls. Chondrification occurs in the fourth month pc from four centers in the adjacent mesoderm. Growth of the cartilaginous portion of the tube is greatest between 16 and 28 weeks pc. Thereafter, increase in tubal length is primarily in the osseous portion of the tube. The changing location of the opening of the auditory tube reflects the growth of the nasopharynx. The tubal orifice is inferior to the hard palate in the fetus, is level with it at birth and well above the hard palate in the adult.

### Second Pharyngeal Pouch

The ventral portion of this pouch is obliterated by the developing tongue. The dorsal portion of this pouch persists in an attenuated form as the tonsillar fossa, the endodermal lining of which covers the underlying mesodermal lymphatic tissue to form the *palatine tonsil.**

### Third Pharyngeal Pouch

The ventral diverticulum endoderm proliferates and migrates from each side to form two elongated diverticula that grow caudally into the surrounding mesenchyme to form the elements of the *thymus gland.* The two thymic rudiments meet in the midline but do not fuse, being united by connective tissue. Lymphoid cells invade the thymus from hemopoietic tissue during the third month pc.

The dorsal diverticulum endoderm differentiates, and migrates caudally to form the *inferior parathyroid gland* (parathyroid III). EYA1 and SIX1/SIX4 genes are essential in patterning the parathyroid and thymus glands. GCM2 is the parathyroid-specific gene and FOXN1 specifies the thymus. The glands derived from the endodermal lining of the pouch lose their connection with the pharyngeal wall when the pouches become obliterated during later development. The lateral glossoepiglottic fold represents the third pharyngeal pouch.

### Fourth Pharyngeal Pouch

The fate of the endoderm of the ventral diverticulum is uncertain: the lining membrane may contribute to thymus or thyroid tissue.

The dorsal diverticulum endoderm differentiates into the *superior parathyroid gland* (parathyroid IV), which, after losing contact with the pharynx, migrates caudally with the thyroid gland. The fetal thyroid gland starts secreting thyroid hormones at 10–12 weeks pc, and responds to its own pituitary gland by 20 weeks pc. The aryepiglottic fold represents the fourth pharyngeal pouch.

### Fifth Pharyngeal Pouch

The attenuated fifth pharyngeal pouch appears as a diverticulum of the fourth pouch. The endoderm of the fifth pouch forms the *ultimopharyngeal*

---

*It should be noted that neither the pharyngeal nor the lingual tonsils originate from the pharyngeal pouch.

*body*. The calcitonin-secreting cells of this structure, however, are derived from neural crest tissue and are eventually incorporated into the thyroid gland. The laryngeal ventricles could represent the remnants of the fifth pharyngeal pouches.

## ANOMALIES OF DEVELOPMENT

Defective development of the pharyngeal pouches results in defects of their derivatives (see Fig. 5–4). The commonest anomalies are pharyngeal fistulae and cysts and persistent tracks of migrated glands derived from the pouches. Atresia of the auditory tube is rare. Congenital absence of the thymus and parathyroid glands, and thus of their products, results in metabolic defects and increased susceptibility to infection (DiGeorge syndrome). DiGeorge syndrome (velocardiofacial syndrome) arises from interstitial deletions of the 22q11.2 region containing the TBX1 gene that is the most prominent participant of the development of conotruncal heart defects, cleft palate, nasal speech and hypocalcemia. Other syndromes affecting pharyngeal arch development include the oculoauriculovertebral spectrum, also known as Goldenhar syndrome, hemifacial microsomia, first- and second-arch syndrome, and craniofacial microsomia. A sporadic and variable group of anomalies that affect development of the mandible, ears and oral structures generally causing facial asymmetry are included in the spectrum of pharyngeal dysmorphology.

## SELECTED BIBLIOGRAPHY

Atlan G, Egerszegi EP, Brochu P, et al. Cervical chondrocutaneous branchial remnants. Plast Reconst Surg 1997; 100:32–39.

Begbie J, Brunet JF, Rubenstein JL, Graham A. Induction of the epibranchial placodes. Development 1999; 126:895–902.

Hickey SA, Scott GA, Traub P. Defects of the first branchial cleft. J Laryng Otol 1994; 108:240–243.

Johnson IJ, Soames JV, Birchall JP. Fourth branchial arch fistula. J Laryng Otol 1996; 110:391–393.

Nofsinger YC, Tom LW, LaRossa D, et al. Periauricular cysts and sinuses. Laryngoscope 1997; 107:883–887.

Scheuerle AE, Good RA, Habal MB. Involvement of the thymus and cellular immune system in craniofacial malformation syndromes. J Craniofac Surg 1990; 1:88–90.

Sperber GH, Sperber SM. Pharyngogenesis. J Dent Assoc South Africa 1996; 51:777–782.

Vade A, Griffiths A, Hotaling A, et al. Thymopharyngeal duct cyst: MR imaging of a third branchial arch anomaly in a neonate. J Magnet Reson Imag 1994; 4:614–616.

Waldhausen JHT, Branchial cleft and arch anomalies in children. Sem Ped Surg 2006; 15:64-89.

Zou D, Silvius D, Davenport J. et al. Patterning of the third pharyngeal pouch into thymus/parathyroid by six and Eyal. Dev Biol 2006; 293:499–512.

# 6 Bone Development and Growth

Bone is formed by two methods of differentiation of mesenchymal tissue that may be of either mesodermal or ectomesenchymal (neural crest) origin. The two varieties of ossification are described as *intramembranous* and *endochondral*. In both, however, the fundamental laying down of osteoid matrix by osteoblasts and its calcification by amorphous and crystalline apatite deposition is similar.

Intramembranous ossification occurs in sheet-like osteogenic membranes, whereas endochondral ossification occurs in hyaline cartilage prototype models of the future bone. The adult structure of osseous tissue formed by the two methods is indistinguishable; furthermore, both methods can participate in forming what may eventually become a single bone, with the distinctions of its different origins being effaced. In the main, the long bones of the limbs and the bones of the thoracic cage and cranial base are of endochondral origin, whereas those of the vault of the skull, the mandible and clavicle are predominantly of intramembranous origin. Membrane bones appear to be of neural crest origin, and arise after the ectomesenchyme interacts with an epithelium.

Control over initiation of osteogenesis resides not in the preosteogenic cells but in adjacent tissues with which they interact inductively. Two genes, namely, core-binding factor alpha CBFA1 (RUNX2/OSF2) and Indian hedgehog (IHH), have been shown to control osteoblast differentiation. Bone morphogenetic proteins (BMPs), members of the transforming growth factor β (TGF-β) superfamily, induce bone formation at genetically designated sites (ossification centers). An intact insulin-like growth factor (IGF)–induced P13-kinase Akt signaling cascade is essential for BMP2-activated osteoblast differentiation and maturation and subsequent bone development and growth. Superimposed on the regulation by transcription and growth factors is endocrine regulation of osteoblasts and osteoclasts. Differentiation of osteoblasts from mesenchyme precedes osteoclast differentiation. Bone deposition by osteoblasts is constantly counterbalanced by osteoclastic bone resorption in a process called *bone remodeling*, regulated by counteracting hypercalcemic parathyroid hormone and hypocalcemic calcitonin. Superimposed upon these two hormones is control exerted by the sex steroid hormones, particularly estrogens. Delayed onset of osteogenesis will reduce the final size of a bone, and premature onset of osteogenesis will increase its final size. Ontogenetically, such timing variation expresses a person's ultimate size range.

During the seventh week pc, mesenchymal cells condense as a prelude to both intramembranous and endochondral bone formation. In the former, cells differentiate into osteoblasts that induce osteoid matrix, thereby forming a center of ossification. In endochondral ossification, the condensed mesenchymal cells initially form a cartilaginous matrix of glycoproteins; this creates a cartilage model of the future bone, and subsequently the osteoid matrix, once it has mineralized, forms a collar of periosteal bone surrounding the cartilage. Invasion and replacement of the cartilage model by bone completes endochondral ossification. Growth of endochondral bones depends upon expansion of retained cartilage (epiphyses) and conversion into bone.

Endochondral bone is three-dimensional in its growth pattern, ossifying from one or more deeply seated and slowly expanding centers. The capacity of cartilage for interstitial growth expansion allows for directed prototype cartilage growth; the cartilage template is then replaced by endochondral bone, accounting for indirect bone growth. Intramembranous bone growth, by contrast, is by direct deposition of osseous tissue in osteogenic (periosteal) membranes; this creates accretional growth, often with great speed, especially over rapidly growing areas, such as the frontal lobes of the brain, or at fracture sites.

Certain inherited congenital defects of bone formation are confined to one type of ossification. Thus, achondroplasia, caused by mutations in the FGFR3 gene, affects only bones of endochondral origin, whereas cleidocranial dysostosis, a delay in the closing of the cranial sutures arising from mutations in the CFBA1/RUNX2 gene, afflicts only those bones of intramembranous origin. In contrast, the inherited condition of osteogenesis imperfecta, caused by mutations in the type I collagen genes (COL1A1, COL1A2), a major component of bone extracellular matrix, afflicts the whole skeleton, whether of endochondral or intramembranous origin.

Ossification starts* at definable points in membranes or cartilages, and from these centers of ossification radiates into the precursor membrane or cartilage. Secondary cartilages, not part of the cartilaginous primordium of the embryo, appear in certain membrane bones (mandible, clavicle) after the onset of intramembranous ossification. Endochondral ossification occurs later in these secondary cartilages of intramembranous bone. The distinction between intramembranous and endochondral bone, although useful at the embryological level of osteogenesis, tends to become insignificant in postnatal life. For example, during repair of fracture to intramembranous bone, cartilage may appear in the healing callus, thereby contradicting its embryological origins. Further, the subsequent remodeling of initially endochondral bones by surface resorption and deposition by the membranous periosteum or endosteum replaces most endochondral bone with intramembranous bone. Consequently, most endochondral bones become a blend of endochondral and intramembranous components.

In the fetus or postnatally, a *primary center of ossification* is the first to appear. This may be followed by one or more *secondary centers*, all of which coalesce into a single bone. Most primary ossification centers appear before birth, whereas secondary centers appear postnatally. Skeletal growth

---

*An accurate complete timetable of the onset of prenatal ossification of all bones is still not available.

is interrupted during the neonatal period, an event that accounts for the natal growth-arrest line in infant bones and teeth. There is evidence of an osteogenesis-inhibiting mechanism in embryonic sutural tissue, accounting for the development of discrete skull bones. Separate bones may fuse into a single composite bone, either as a phylogenetic phenomenon, exemplified in the cranial base, or as an ontogenetic phenomenon, exemplified in the fusion of several bones of the skull calvaria into a single bone in extreme old age.

Details of the mechanisms of ossification and the processes of calcification, deposition and resorption of bone through the operation of osteoblasts and osteoclasts are best studied in histology and physiology texts. The dependence of osseous tissue upon the metabolism of calcium and phosphates means that the structure of bone is a sensitive indicator of the state of turnover of these and other bone minerals. Moreover, bone growth, maintenance, repair and degeneration depend upon the actions of certain hormones and vitamins, which operate indirectly through control of calcium and phosphate and general metabolism or directly by their varying influences on growth cartilages. Appreciation of these complex physiological and biochemical phenomena is necessary for understanding the mechanisms of morphogenesis of adult bone structure and shape, but a discussion of this is outside the scope of the present work.*

The basic shape of bones, and to a considerable degree their size, are genetically determined. Once this inherited morphology is established, environmentally variable minor features of bones, such as ridges, and so on, develop. Superimposed upon the basic bone architecture are nutritional, hormonal and functional influences that, because of the slow and continual replacement of osseous tissue throughout life, enable bones to respond morphologically to functional stresses. Specific periosteal and capsular functional matrices influence specific portions of related bones, termed *skeletal units*. Macroskeletal units may consist of a single bone or of adjacent portions of several bones (e.g., the frontal, parietal and temporal bones of the calvaria). Each macroskeletal unit is made up of microskeletal units that respond independently to functional matrices and thereby determine the varying shapes of the macroskeletal unit or classically named bone. Although the specific growth rates of the individual microskeletal units might differ, there is nonetheless a constant proportionality between growth rates, thereby imparting a fairly constant shape to the enlarging macroskeletal unit.

Based upon the influence of muscles on bone morphology, three classes of morphological features of the craniofacial skeleton have been identified: (1) those that never appear unless muscles are present (e.g., temporal line, nuchal lines); (2) those that are self-differentiating but

---

*An additional factor of significance in bone physiology and morphology is its hemopoietic function. Bone marrow hemopoiesis begins in the third month pc, and rapidly replaces liver hemopoiesis as the chief site of blood cell formation.

require muscles to persist (e.g., angular process of mandible); (3) those that are largely independent of the muscles with which they are associated (e.g., the zygomatic bone or the body of the mandible).

The exact mechanism by which functional deforming mechanical forces produce structural bone changes is obscure. One theory postulates mediation of biomechanical stresses through piezoelectric currents created by bioelectrical factors. Bone, whose collagen and apatite content renders it highly crystalline, behaves as a crystal when it is mechanically deformed: it generates a minute electric current, thereby producing polar electric fields. Bioelectric effects may be generated in several cell membranes. It is conceivable that osteoclasts and osteoblasts and the matrix within which they operate react to electric potentials by building up bone (experiments suggest) in negatively charged fields, and, conversely, resorbing bone in positively charged fields. These stress potential currents may allow for the adjustments in bone structure that are made to meet new functional demands.

Another theory of bone remodeling postulates a mechanochemical hypothesis whereby mechanical stresses are translated into osteoblastic/osteoclastic activity. A change in bone loading results in altered straining of hydroxyapatite crystals that alters their solubility, changing local calcium activity that either stimulates or resorbs bone.

In orthodontics, distinction is made between genetically determined unalterable "basal" bone and the superimposed "functional" bone that is amenable to manipulated alteration. In practice, "basal" bone refers to the body of the maxilla or mandible, and "functional" bone is the alveolar bone of both jaws that supports the teeth and responds to orthodontic forces. The ultimate shape of a bone and its internal architecture, then, is a reflection of its inherited form and the mechanical demands to which it is subjected. Intrinsic genetic factors may play only an initial role in determining the size, shape and growth of a bone. Extrinsic functional or environmental factors become the predominant determinator of bone form. Because the environment is constantly changing, bones never attain a "final" morphology and their shapes are continually subject to change. It has been observed that a bone is composed of the minimal quantity of osseous tissue that will withstand the usual functional stresses applied to it. This supposedly accounts for the hollow marrow centers of long bones and possibly the sinuses in the skull bones. These factors were formulated into a trajectorial theory of bone structure by Culmann and Meyer just after the middle of the 19th century and proposed as Wolff's Law (1870), which states that changes in the function of a bone are attended by alterations in its structure.

During adolescence there is a spurt in bone growth that is believed to be mediated by circulating growth hormones. Bones may differ in the timing of their maximal increases within individuals, suggesting that intrinsic factors in different skull bones may be important in determining changes in the rate of growth at different ages. Circulating growth hormones alone do not determine the timing of the adolescent growth spurt. The pattern of timing of variations in bone growth velocity is intrinsically, and presumably genetically, determined. Hormones augment a genetically regulated pattern of growth rate in the cranial base.

An interposed cartilage (that is, an epiphysis)* converting to osseous tissue adds to an endochondral bone's length and simultaneously displaces adjacent parts of the bone by expansion in opposite directions. Embryonic cartilage cells are arranged haphazardly, precluding directionality of growth. By contrast, specialized (epiphyseal) growth plates contain organized columns of cartilage cells, accounting for highly directed growth. This direction-oriented growth is based upon chondrocytes being stacked in coin-like columns and their proliferation acting as hydraulic jacks. Cartilage can grow under weight pressure because of its avascularity; its nutrition is provided by perfusing tissue fluids that are not obstructed by load pressures. The growth forces originate for the most part within the bone's cartilage. On the other hand, most intramembranous bones with sutural contacts become separated by external, capsular functional matrix growth forces (e.g., the expanding brain or eyeball). Addition of osseous tissue to these membranous sutural surfaces passively fills in the widened interval in a field of tension, contrasting with the compression field created by the endochondral mechanism.

The overall growth of bones, resulting in their recognizable expansion, is a function of two phenomena: namely, remodeling and transposition. Remodeling is a combination of accretional growth and resorption of bone and is a response, at least in part, to periosteal functional matrices. Because of the rigid nature of calcified bone, growth of this tissue must be appositional, by cortical surface deposition of newly formed bone. This primary mode of bone growth is in contrast to the interstitial form of growth that takes place in most soft tissues, which can expand in size by division and growth of cells within the tissue (Fig. 6–1). Such internal expansion is not possible in rigid bone. Concomitant with the deposition of bone in certain areas, resorption occurs in other areas to allow for remodeling; that is, changing the shape of a bone. The periosteum covering the surface of bone provides a ready source of osteoblasts for deposition and osteoclasts for resorption. The rate of remodeling diminishes when growth slows down, and bone density increases.

The second basic phenomenon of bone growth, that is, transposition, is displacement of the remodeling bones vis-à-vis each other. Bone displacement is the result of forces exerted by the surrounding soft tissues (capsular matrices) and by the primary intrinsic growth of the bones themselves. Accordingly, bone growth overall represents the cumulative effects of intrinsic remodeling (a vector of deposition and resorption) and

---

*The bones of teleost fishes do not have cartilaginous growth plates. Appositional growth of their bones has no limit. Amphibians and reptiles possess cartilaginous epiphyses that persist throughout life, providing for potentially continuous growth. The osseous fusion of mammalian epiphyses limits the growth potential of their bones.

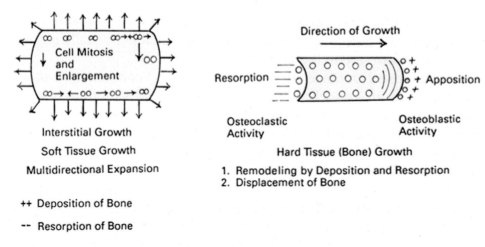

Figure 6–1  Modes of growth.

displacement. These phenomena may occur in the same or diverse directions, but their combination is usually complex and their relative contributions are difficult to determine. This difficulty is the source of much controversy relating to craniofacial bone growth.

## BONE ARTICULATIONS

Because the sites of junction between bones, known as *articulations* or *joints*, are important in relation to bone growth, it is useful to classify them at this juncture (Figure 6–2).

### Movable Joints (Diarthroses)

All movable joints are characterized by a synovial membrane lining a joint cavity. Synovial joints are classified according to the shapes of the participating bones (e.g., ball-and-socket joints) or the nature of their action (e.g., hinge joints).

### Immovable Joints (Synarthroses)

| Junctional tissue | Articulation type | Example |
| --- | --- | --- |
| Fibrous connective tissue | Syndesmosis | Skull suture |
| Cartilage | Synchondrosis | Symphysis pubis |
| Bone | Synostosis | Symphysis menti |

Diarthroses, per se, do not play a significant role in bone growth except in the temporomandibular joint. The articular cartilage of this joint alone, of all the articular cartilages, provides a growth potential for one of the articulating bones; namely, the mandible.

Synarthroses, on the other hand, play a very significant part in the growth of the articulating bones.

The apposition of bone during growth may take two forms:

1.  *Surface deposition*, accounting for increased thickness of bone that may be modified in remodeling by selective resorption.

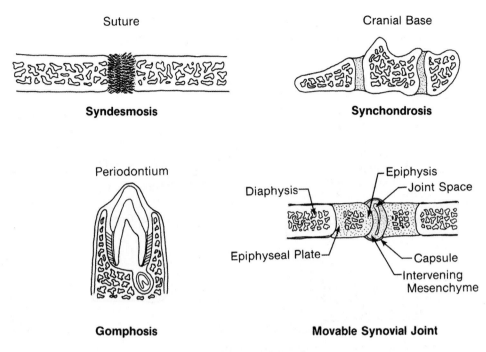

Figure 6–2  Various types of joints.

**2.** Sutural deposition, restricted to the opposing edges of bones at a suture site, and accounting for the "filling in" of expanded sutures as a result of displacement (Figure 6–3).

Both of these methods of bone growth are used in different areas of the skull for its expansion in size and its remodeling. Remodeling of growing bone maintains proportions within and between bones. Sutural planes tend to be aligned at right angles to the direction of movement of growing bones. The bony surfaces are oriented to slide in relation to one another, as the growing bones move apart. Displacement of bones is an important factor in the expansion of the craniofacial skeleton. The sliding characteristic of bone movement at angled sutural surfaces accommodates the need for continued skeletal growth and determines its direction. The basic suture type is the "butt-end," or flat end-to-end, type. Beveling and serrations occur in sutures in response to functional demands. Sutural serrations are a gross manifestation of a form of trabecular growth responsive to tensions within the sutural tissues; in all probability set up by, for example, the rapidly expanding brain, eyeball or nasal septum.

Incremental growth at sutures does not take place only in the plane of the preexisting curved bones. Differential rates of marginal and surface growth at the suture site allow the plane of growth to alter as the bone edges grow. Resorption of surface bone plays a significant role in this remodeling process. Differential apposition of bone leads to greater or lesser growth of the individual sutural bones, each bony margin contributing independently of the other. In this manner, both remodeling and growth can take place at suture sites. The sutural growth potential is used to advantage in the orthodontic treatment of skeletal deficiencies by

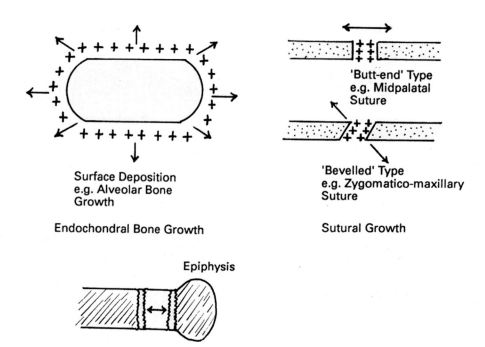

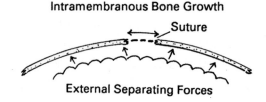

Figure 6–3 Modes of bone growth.

techniques of forced expansion, as in intermaxillary expansion of narrow palates.

Sutural fusion, by ossification of the syndesmotic articulation, indicates cessation of growth at that suture. There is great variability in the timing of sutural closures, making this an unreliable criterion of age. Premature synostosis reduces the cranial diameter at right angles to the fused suture, and abnormal compensatory growth in other directions results in malformation of the skull (e.g., *scaphocephaly*, wedge-shaped cranium; *acrocephaly* or *plagiocephaly*, pointed or twisted cranium) (see Chapter 14).

## SELECTED BIBLIOGRAPHY

Bais MV, Wigner N, Young M et al. BMP2 is essential for post natal osteogenesis but not for recruitment of osteogenic stem cells. Bone 2009; 45:254–266.

Cohen MM Jr. The new bone biology: pathologic, molecular and clinical correlates. Am J Med Genet Part A 2006; 140A:2646–2706.

Conen KL, Nishimori S, Provot S, Kronenberg HM. The transcriptional cofactor Lbh regulates angiogenesis and endochondral bone formation during fetal bone development. Dev Biol 2009; 333:348–358.

Delezoide AL, Benoist-Lasselin C, Legeai-Mallet L, et al. Spatio-temporal expression of FGFR 1, 2 and 3 genes during human embryo-fetal ossification. Mech Dev 1998; 77:19–30.

Goode J (ed). The molecular basis of skeletogenesis. Novartis Foundation Vol 232. Chichester. UK: John Wiley & Sons 2000.

Hall BK. Bone: A treatise. Vols I–VII. Boca Raton, FL: CRC Press, 1991.

Hillarby MC, King KE, Brady G, et al. Localization of gene expression during endochondral ossification. Ann N Y Acad Sci 1996; 785:263–266.

Jonason JH, Xiao G, Zhang M, Xing L, Chen D. Post-translational regulation of Runx2 in bone and cartilage. J Dent Res 2009; 88:693–703.

Karsenty G. The genetic transformation of bone biology. Genes Dev 1999; 13:3037–3051.

Liu F, Malaval L, Aubin JE. The mature osteoblast phenotype is characterized by extensive plasticity. Exp Cell Res 1997; 232:97–105.

Mukherjee A, Rotwein P. Akt promotes BMP2-mediated osteoblast differentiation and bone development. J Cell Sci 2009; 122:716–726.

Thorogood P, Sarkar S, Moore R: Skeletogenesis in the head. In Oral Biology at the turn of the century. Guggenheim B, Shapiro S, (eds). Basel: S. Karger AG 1998:93–100.

Wezeman FH. Morphological foundations of precartilage development in mesenchyme. Microsc Res Tech 1998; 43:91–101.

# SECTION II

---

# CRANIOFACIAL DEVELOPMENT

*In the closest union there is still some separate existence of component parts; in the most complete separation there is still a reminiscence of union.*

The Notebooks of Samuel Butler

# INTRODUCTION

Development of the skull, comprising both the cranium and the mandible, is a blend of the morphogenesis and growth of three main skull entities (see Figure) arising from neural crest and paraxial mesoderm tissues. These skull entities are composed of:

1. The Neurocranium
   - The *Vault of the Skull* or *Calvaria*: Phylogenetically of recent origin to cover the newly expanded brain, it is formed from intramembranous bone of paraxial mesodermal origin and is known as the *desmocranium* (*desmos*, membrane).
   - The *cranial base*: Derived from the phylogenetically ancient cranial floor, it is associated with the capsular investments of the nasal and auditory sense organs; formed from endochondral bone of neural crest origin, its cartilaginous precursor is known as the *chondrocranium* (*chondros*, cartilage).
2. The Face (*orognathofacial complex*): Derived from modifications of the phylogenetically ancient branchial arch structures; formed from intramembranous bone of neural crest origin; also known as the *splanchnocranium* (*splanchnos*, viscus) or *viscerocranium* (*viscus*, an organ). This complex forms the visual, aural, respiratory aditus and the oromasticatory apparatus.
3. The Masticatory Apparatus: comprised of the *jaw bones*, their joints and musculature and the teeth. The Dentition: derived phylogenetically from ectodermal placoid scales, reflected in the embryological development of the teeth from oral ectoderm (dental lamina) and neural crest (dental papilla).

Embryonic origins of the skull

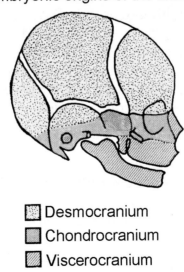

- Desmocranium
- Chondrocranium
- Viscerocranium

Main developmental divisions of the skull.

The cranial base is, to some extent, shared by both neurocranial and facial elements. The masticatory apparatus is composed of both facial and dental elements. The skull is thus a mosaic of individual components, each of which enlarges during growth in the proper amount and direction to attain and maintain the stability of the whole.

Each of the three main craniofacial entities possesses different characteristics of growth, development, maturation and function. Yet each unit is so integrated with the others that coordination of the growth of all is required for normal development. Failure of correlation of the differing growth patterns, or aberration of inception or growth of an individual component, results in distorted craniofacial relationships and is a factor in the development of dental malocclusion. Although both the neurocranium and the face have mixed intramembranous and endochondral types of bone formation, the bony elements of the masticatory apparatus are predominantly of intramembranous origin. The dental tissues have an ectodermal origin for their enamel and a neural crest origin for much of the mesenchyme that forms the dentine, pulp, cementum and periodontal ligament.

Interestingly, the historically more recent developments in the mammalian skull, namely, the membrane bones of the jaws and facial skeleton (splanchnocranium), are more susceptible to developmental anomalies than are the older cartilaginous parts of the skull (chondrocranium). Developmental defects of the face and jaws are relatively common, whereas congenital defects of the skull base and the nasal and otic (auditory) capsules are relatively rare. During postnatal growth of congenital craniofacial defects, three general patterns of development have been observed. Hypoplastic defects may improve, with "catch-up" growth minimizing the defect. Alternatively, the defective pattern of growth is maintained throughout infancy or childhood so that the malformation is retained to the same degree in the adult. The third pattern is one in which the developmental derangement worsens with age, the severity of the malformation becoming greater in adulthood.

Differentiation and growth of the chondrocranium appears to be strongly genetically determined and subject to minimal environmental influence. Diseases of defective endochondral bone formation are reflected in abnormalities of the skull base. On the other hand, growth of the desmocranium and splanchnocranium appears to be subject to minimal genetic determination but is strongly influenced by local environmental factors.

The cranial components surrounding the special sense organs of olfaction, sight, hearing and balance are almost full grown at birth. The remaining cranial elements grow and change considerably postnatally in keeping with enlargement and usage of adjacent structures. The calvaria grows most rapidly in response to the early expanding brain, followed by the nasal airway system determining midfacial development. The masticatory system is the last major functional system to reach maturity.

For descriptive convenience, details of the growth and development of the head have been subdivided into the following components:

Calvaria
Cranial base
Facial skeleton
Palate
Paranasal sinuses
Mandible
Temporomandibular joint
Skull growth: sutures and cephalometrics
Tongue and tonsils
Salivary glands
Muscle development
Special sense organs
Dentition

# 7 Calvaria

## MEMBRANOUS NEUROCRANIUM (DESMOCRANIUM)

The mesenchyme that gives rise to the vault of the neurocranium is first arranged as a capsular membrane around the developing brain. The membrane is composed of two layers: an inner *endomeninx*, primarily of neural crest origin, and an outer *ectomeninx*, of mixed paraxial mesodermal and neural crest origin (Fig. 7–1). The endomeninx forms the two leptomeningeal coverings of the brain—the pia mater and the arachnoid. The ectomeninx differentiates into the inner dura mater covering the brain, which remains unossified, and an outer superficial membrane with chondrogenic and osteogenic properties. Osteogenesis of the ectomeninx occurs as intramembranous bone formation over the expanding dome of the brain, forming the skull vault, or calvaria, whereas the ectomeninx forming the floor of the brain chondrifies as the chondrocranium, which later ossifies endochondrally (Fig. 7–2).

Despite their divergent fates, the two layers of the ectomeninx remain in close apposition except in regions where the venous sinuses develop. The dura mater and its septa, the falces cerebri and cerebelli and the tentorium cerebelli, show distinctly organized fiber bundles closely related and strongly attached to the sutural systems that later develop in the vault. The adult form of the neurocranium is the end result of the preferential direction of the forces set up by growth of the brain constrained by these dural fiber systems. Without the dural bands, the brain would expand as a perfect sphere. Because the dura mater serves as the endocranial periosteum, it also determines the shape of the calvarial bones.

In the somite-period embryo, the neural tube's covering dura mater and its surface ectoderm are in contact in the area of the closing anterior neuropore of the developing brain. Transient maintenance of this contact during development causes a dural projection that, consequent upon ventral bending of the rostrum, extends into the future frontonasal suture area. Later, as the nasal capsules surround the dural projection, the resulting midline canal forms the basis of the foramen cecum where the ethmoid-frontal bone junction develops. The dural projection and frontonasal area skin normally separate, allowing the canal to close, forming

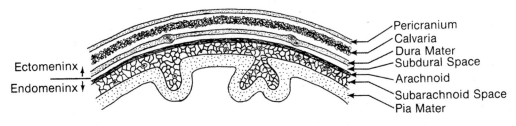

**Figure 7–1** Neurocranial derivatives of the embryonic ectomeninx (calvaria, dura mater) and endomeninx (arachnoid and pia mater).

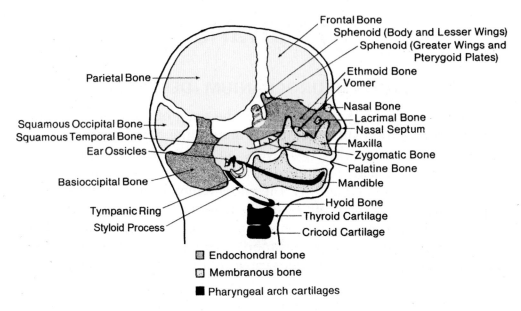

Figure 7–2  Skull bone origins.

the *foramen cecum* (hence "blind foramen"). If the foramen fails to close, this leaves an abnormal pathway for neural tissue to herniate into the nasal region. Herniated portions of the brain are known as *encephaloceles*, congenital anomalies most commonly seen in the frontal bone and occipital regions (Fig. 7–3).

Ossification of the intramembranous calvarial bones depends upon the presence of the brain; in its absence (anencephaly), no bony calvaria forms (see Fig. 3–18). Several primary and secondary ossification centers develop in the outer layer of the ectomeninx to form individual bones. The mesodermally derived ectomeninx gives rise to major portions of the frontal, parietal, sphenoid, petrous temporal and occipital bones (Fig. 7–4). The neural crest provides the mesenchyme forming the lacrimal, nasal,

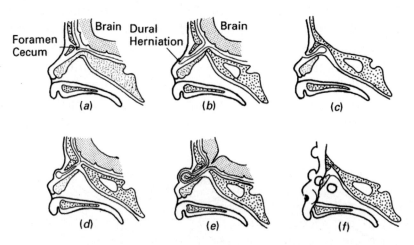

Figure 7–3  Patterns of meningoencephalocele herniation through the foramen cecum forming facial defects. (a) Embryonic stage, dura within foramen cecum. (b) During fetal development, dura herniates through foramen cecum and contacts skin. Normally retracts before birth. (c) Dermoid sinus with cyst. (d) Dermoid cyst, may or may not have stalk to dura. (e) Encephalocele. (f) Possible sites of defects.

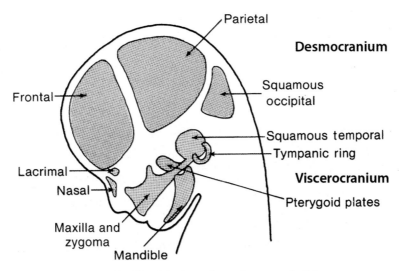

**Figure 7–4** Ossification sites of membranous skull bones.

squamous temporal, zygomatic, maxillary and mandibular bones. These embryonic distinctions of bone derivation account for different syndromic characteristics of bone lesions in neurocristopathies versus mesodermal-based pathologies.

A pair of *frontal bones* appears from single primary ossification centers forming in the region of each superciliary arch at the eighth week post conception (pc). Three pairs of secondary centers appear later—in the zygomatic processes, nasal spine and trochlear fossae. Fusion between these centers is complete at 6 to 7 months pc. At birth, the frontal bones are separated by the frontal (metopic) suture; synostotic fusion of this suture usually starts about the second year and unites the frontal bones into a single bone by 7 years of age.*

The two parietal bones arise from two primary ossification centers for each bone which appear at the parietal eminence in the eighth week pc and fuse in the fourth month pc. Delayed ossification in the region of the parietal foramina may result in a sagittal fontanelle at birth.

The supranuchal *squamous portion of the occipital bone* (above the superior nuchal line) ossifies intramembranously from two centers, one on each side, appearing in the eighth week pc. The rest of the occipital bone ossifies endochondrally (see p. 110).

The *squamous portion of the temporal bone*[†] ossifies intramembranously from a single center appearing at the root of the zygoma at the eighth week pc. The *tympanic ring of the temporal bone* ossifies intramembranously from four centers, appearing in the third month pc in the lateral wall of the tympanum. The two membranous bone portions of the temporal bone fuse at birth. The rest of the temporal bone ossifies endochondrally (see p. 111).

---

*The frontal (metopic) suture persists into adulthood in 10% to 15% of skulls. In such cases, the frontal sinuses are absent or hypoplastic.

[†]The squamous temporal bone is independent of brain induction and is present in anencephaly.

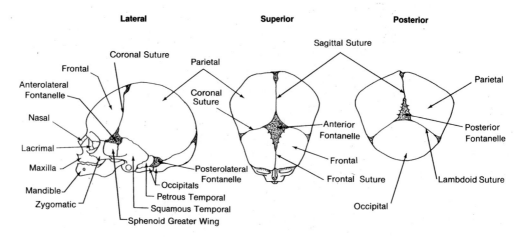

Figure 7–5  Fontanelles and sutures of the calvaria.

Should any unusual ossification centers develop between individual calvarial bones, their independent existence is recognizable as small *sutural* or *wormian bones*.[‡] The earliest centers of ossification first appear during the seventh and eighth weeks pc, but ossification is not completed until well after birth. The mesenchyme between the bones develops fibers to form syndesmotic articulations. The membranous mesenchyme covering the bones forms the periosteum. At birth, the individual calvarial bones are separated by sutures of variable width and by fontanelles. Six of these fontanelles are identified as the anterior, posterior, posterolateral and anterolateral, in relation to the corners of the two parietal bones (Fig. 7–5). These flexible membranous junctions between the calvarial bones allow narrowing of the sutures and fontanelles and overriding of these bones when they become compressed during birth in traversing the pelvic birth canal. The head may appear distorted for several days after birth (Fig. 7–6).

Postnatal bone growth results in narrowing of the sutures and elimination of the fontanelles. The anterolateral fontanelles[*] close 3 months after birth; the posterolateral ones close during the second year; the posterior fontanelle closes 2 months after birth, and the anterior one during the second year. The median frontal suture is usually obliterated between 6 and 8 years of age. Fusion of the frontal (metopic) suture involves chondroid tissue, occasionally forming secondary cartilage, which is progressively substituted by lamellar bone. This extension of ossification of the calvarial bones continues throughout life: the syndesmosal sutures between the

[‡] Wormian bones occur most frequently along the lambdoid suture, where they form inter parietal bones. Development of these ossicles may have a genetic component, but as nearly all hydrocephalic crania have wormian bones, it appears that deforming stress is a contributing factor.

[*] The site of the anterolateral fontanelle corresponds to pterion in the adult skull; the site of the posterolateral fontanelle corresponds to asterion.

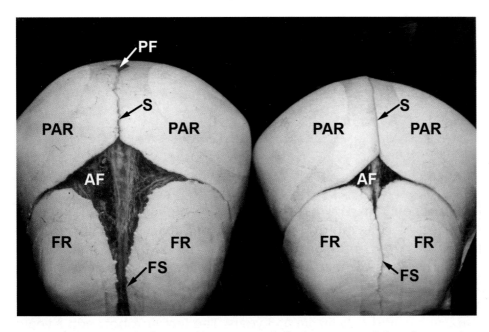

**Figure 7–6** Top view of calvariae of two neonate skulls depicting variability of anterior fontanelle size at birth. AF, Anterior fontanelle; FR, frontal bone; FS, frontal (metopic) suture; PAR, parietal bone; PF, posterior fontanelle; S, sagittal suture.

neurocranial bones fuse into synostoses with advancing years, uniting the individual calvarial bones into a single component in old age. In the fetus, the intramembranous neurocranial bones are molded into large, slightly curved "plates" over the expanding brain which they cover.

The precocious development of the brain determines the early predominance of the neurocranium over the facial and masticatory portions of the skull. Although the brain, and consequently the neurocranial bone vault, develop very rapidly very early, their growth slows and ceases at an earlier age than do the facial and masticatory elements of the skull. The predominance of the neurocranium over the face is greatest in the early fetus, reducing to an 8 to 1 proportion at birth, 6 to 1 in the second year, 4 to 1 in the fifth year and a 2 to 2.5 to 1 proportion in the adult. At birth, the neurocranium has achieved 25% of its ultimate growth; it completes 50% by age 6 months and 75% by 2 years. By 10 years of age, neurocranial growth is 95% complete, but the facial skeleton has achieved only 65% of its total growth. In postnatal life, the neurocranium increases 4 to 5 times in volume, whereas the facial portion increases some 8 to 10 times its volume at birth.

The ultimate shape and size of the cranial vault are primarily dependent upon the internal pressures exerted on the inner table of the neurocranial bones. The expanding brain exerts separating tensional forces upon the bone sutures, thereby secondarily stimulating compensatory sutural bone growth (Fig. 7–7). The brain acts in this context as a "functional matrix" in determining the extent of neurocranial bone growth. The circumference of the head, because it is related to intracranial volume, is a

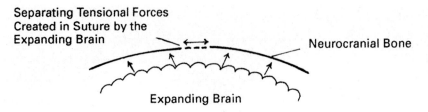

**Figure 7–7** Schematic section through skull and brain to demonstrate the forces of "functional matrix" growth.

good indicator of brain growth.* The precocious early development of the brain is reflected in the rapidly enlarging circumference of the head, which nearly doubles from an average of 18 cm at the midgestational period (4 to 5 months) to an average of 33 cm (approximately 13 inches) at birth. This rapid increase of head circumference continues during the first year, reaching an average of 46 cm, and then slows; head circumference reaches 49 cm at 2 years, and only 50 cm at 3 years. The increase between 3 years and adulthood is only about 6 cm.

Growth of the calvarial bones is a combination of (1) sutural growth, (2) surface apposition and resorption (remodeling) and (3) centrifugal displacement by the expanding brain. The proportions attributable to the various growth mechanisms vary. Accretion to the calvarial bones is predominantly sutural until about the fourth year of life, after which surface apposition becomes increasingly important. Remodeling of the curved bony plates allows for their flattening out to accommodate the increasing surface area of the growing brain. The flattening of the early high curvature of the calvarial bones is achieved by a combination of endocranial erosion and ectocranial deposition, together with ectocranial resorption from certain areas of maximal curvature such as the frontal and parietal eminences.

The bones of the newborn calvaria are unilaminar and lack diploë. From about 4 years of age, lamellar compaction of cancellous trabeculae forms the inner and outer tables of the cranial bones. The tables become continuously more distinct into adulthood. This differential bone structure creates a high stiffness to weight ratio, with no relative increase in the mineral content of cranial bone from birth to adulthood. Whereas the behavior of the inner table is related primarily to the brain and intracranial pressures, the outer table is more responsive to extracranial muscular and buttressing forces. However, the two cortical plates are not completely independent. The thickening of the frontal bone in the midline at the gla-

---

*Between birth and adulthood, brain size increases about 3.5 times, from approximately 400 cm³ to 1300 to1400 cm³. The early rapid expansion of the brain poses problems for the already established cerebral blood vessels: it stretches them, weakening their walls, predisposing them to aneurysm formation in later life.

bella results from separation of the inner and outer tables with invasion of the frontal sinus between the cortical plates. Only the external plate is remodeled, as the internal plate becomes stable at 6 and 7 years of age, reflecting the near cessation of cerebral growth. Thus, only the inner aspect of the frontal bone can be used as a stable (x-ray) reference point for growth studies from age 7 years onward. Growth of the external plate during childhood produces the superciliary arches, mastoid processes, external occipital protuberance and temporal and nuchal lines that are all absent from the neonatal skull. The bones of the calvaria continue to thicken slowly even after their general growth is complete.

When intracranial pressures become excessive, as in hydrocephalus, both plates of the bones of the calvaria become thinned and grossly expanded. Conversely, the reduced functional matrix force of the brain in microcephalics results in a small calvaria. Normal forces acting on the outer table of bone alone tend to influence the superstructure of the cranium only, and not the intracranial form. The pull of muscles would account to some degree for the development of the mastoid process, lateral pterygoid plate, temporal and nuchal lines in the cranium, coronoid process and ramus-body angle of the mandible. In the face, the buttressing resistance to masticatory forces produces the supraorbital processes, superstructural bony projections that add to the dimension of the cranium, but are unrelated to the intracranial capacity. Abnormal external forces applied during development can distort cranial morphology, but, strangely enough, not cranial capacity, as is evident by the bizarre shapes of skulls produced by pressure devices on children's skulls in some tribal societies. These artifically shaped skulls are named according to their distortions, acrocephaly, platycephaly, brachycephaly, and so forth. (Fig. 7–8).

## ANOMALIES OF DEVELOPMENT

The calvaria is particularly susceptible to a number of congenital defects, ranging from chromosomal to hormonal in their etiology. The time of closure of the sutures is altered in many of these afflictions, leading to variable distortions of skull shape. In such widely different conditions as cretinism, progeria, trisomy 21 and cleidocranial dysostosis, there is delayed midline ossification of the frontal (metopic) and sagittal sutures of the calvaria, so that the anterior fontanelle may remain open into adult life. The resulting brachycephalic skull results in a "bossed" forehead of highly curved frontal and parietal bones and hypertelorism, partly obscuring the smaller brain case.

Premature fusion of sutures (craniosynostosis) is dealt with in Chapter 14 (p. 172). Defects in closure of the foramen cecum at the ethmoid-frontal suture allow herniation of the cranial contents into the face, forming frontal encephaloceles (see Fig. 7–3). Occipital encephaloceles may occur through cranioschisis (fissured cranium) lesions. Basal encephaloceles protrude through the skull base.

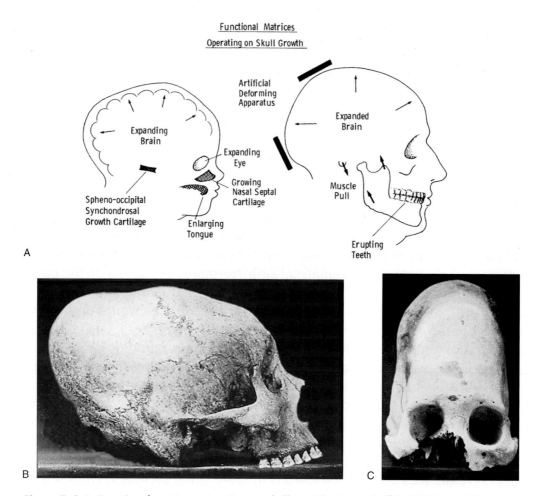

**Figure 7–8** A. Functional matrices operating on skull growth. B. An adult skull distorted into an antero-posterior elongation by head-boards in infancy. C. An adult skull distorted into an acrocephalic shape by constraints in infancy.

## SELECTED BIBLIOGRAPHY

Cho JY, Lee WB, Kim HJ, et al. Bone-related gene profiles in developing calvaria. Gene 2006; 372:71–81.

Evans C, Marton T, Rutter S et al. Cranial vault defects: the description of three cases that illustrate a spectrum of anomalies. Ped Dev Path 2009;12:96–102.

Harris CP, Townsend JJ, Carey JC. Acalvaria: a unique congenital anomaly. Am J Med Genet 1993; 46:694–699.

Iseki S, Wilkie AO, Heath JK, et al. Fgfr2 and osteopontin domains in the developing skull vault are mutually exclusive and can be altered by locally applied FGF2. Development 1997; 124:3375–3384.

Kim HJ, Rice DP, Kettunen PJ, Thesleff I. FGF-, BMP- and SHH-mediated *signaling* pathways in the regulation of cranial suture morphogenesis and calvarial bone development. Development 1998;125:1241–1251.

Liu YH, Kundu R, Wu L, et al. Premature suture closure and ectopic cranial bone in mice expressing Msx2 transgenes in the developing skull. Proc Natl Acad Sci U S A 1995; 92:6137–6141.

Mandarim-de-Lacerda CA, Alves MU. Growth of the cranial bones in human fetuses (2nd and 3rd trimesters). Surg Rad Anat 1992; 14:125–129.

Neumann K, Moegelin A, Temminghoff M, et al. 3D-computed tomography: a new method for the evaluation of fetal cranial morphology. J Craniofac Genet DevBiol 1997; 17:9–22.

Raines C. Primary acalvaria. J Diag Med Sonog 2006; 22:407–410.

Silau AM, Fischer Hansen B, Kjaer I. Normal prenatal development of the human parietal bone and interparietal suture. J Craniofac Genet Dev Biol 1995; 15:81–86.

Sperber GH, Honore LH, Johnson ES. Acalvaria, holoprosencephaly and facial dysmorphia syndrome. J Craniofac Genet Dev Biol 1986; 6 (Suppl. 2): 319-329.

Toma CD, Schaffer JL, Meazzini MC, et al. Developmental restriction of embryonic calvarial cell populations as characterized by their in vitro potential for chondrogenic differentiation. J Bone Mineral Res 1997; 12:2024–2039.

Villaneuva JE, Nimni ME. Modulation of osteogenesis by isolated calvaria cells: a model for tissue interactions. Biomaterials 1990; 11:1921.

# 8 Cranial Base

## CHONDROCRANIUM

During the 4th week postconception (pc), mesenchyme derived from the paraxial mesoderm and neural crest condenses between the developing brain and foregut to form the base of the ectomeningeal capsule. This condensation betokens the earliest evidence of skull formation. Even then, development of the skull starts comparatively late, after development of the primordia of many of the other cranial structures, such as the brain, cranial nerves, eyes and blood vessels. During the late-somite period, the occipital sclerotomal mesenchyme concentrates around the notochord underlying the developing hindbrain. From this region, the mesenchymal concentration extends cephalically, forming a floor for the brain. Conversion of the ectomeninx mesenchyme into cartilage* constitutes the beginning of the chondrocranium, starting on day 40 pc. The anterior (rostral) prechordal region of cranial base is of neural crest origin, whereas the posterior (caudal) chordal region is of mesodermal origin. The boundary between these regions is demarcated at the conjunction of the future basisphenoid and basioccipital bones.

Although the chondrocranium is a continuous cartilaginous structure, it develops from a number of distinct cartilages that appear at specific locations and times. The parachordal cartilages are the first to appear alongside the cranial end of the notochord with the expression of type II collagen and SOX9 transcription factor. The notochord is a major signaling center for patterning the parachordal cartilages (Fig. 8–1). From these parachordal cartilages, a caudal extension of chondrification incorporates the fused sclerotomes arising from the four occipital somites surrounding the neural tube. The sclerotome cartilage, the first part of the skull to develop, forms the boundaries of the foramen magnum, providing the anlagen for the basilar and condylar parts of the occipital bone.

The cranial end of the notochord is at the level of the oropharyngeal membrane which closes off the stomodeum. Just cranial to this membrane, the hypophyseal (Rathke's) pouch arises from the stomodeum; the pouch gives rise to the anterior lobe of the pituitary gland (adenohypophysis), lying immediately cranial to the termination of the notochord (Fig. 8–2; see Fig. 3–1). Two *hypophyseal (postsphenoid) cartilages* develop on either side of the hypophyseal stem and fuse to form the *basisphenoid (postsphenoid)* cartilage, which contains the hypophysis, and will later give rise to the sella turcica and posterior part of the body of the sphenoid bone.

---

*Formation of the cartilages of the chondrocranium is dependent upon the presence of the brain and other neural structures, and an appropriately staged inducing epithelium. Chondrogenesis will occur only after an epithelial-mesenchymal interaction has taken place.

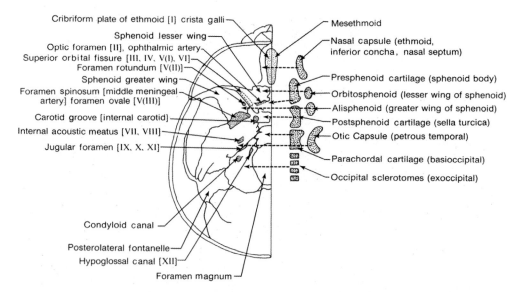

Cribriform plate of ethmoid [I] crista galli
Sphenoid lesser wing
Optic foramen [II], ophthalmic artery
Superior orbital fissure [III, IV, V(I), VI]
Foramen rotundum [V(II)]
Sphenoid greater wing
Foramen spinosum [middle meningeal artery] foramen ovale [V(III)]
Carotid groove [internal carotid]
Internal acoustic meatus [VII, VIII]
Jugular foramen [IX, X, XI]

Mesethmoid
Nasal capsule (ethmoid, inferior concha, nasal septum)
Presphenoid cartilage (sphenoid body)
Orbitosphenoid (lesser wing of sphenoid)
Alisphenoid (greater wing of sphenoid)
Postsphenoid cartilage (sella turcica)
Otic Capsule (petrous temporal)
Parachordal cartilage (basioccipital)
Occipital sclerotomes (exoccipital)

Condyloid canal
Posterolateral fontanelle
Hypoglossal canal [XII]
Foramen magnum

**Figure 8–1** The primordial cartilages of the chondrocranium (right half) and their derivatives (left half). The foramina and their contents, blood vessels and cranial nerves (roman numerals in square brackets) are identified on the left.

Cranial to the pituitary gland, fusion of two *presphenoid* (*trabecular*) cartilages forms the precursor to the presphenoid bone that will form the anterior part of the body of the sphenoid bone. Laterally, the chondrification centers of the *orbitosphenoid* (lesser wing) and *alisphenoid* (greater wing) contribute wings to the sphenoid bone. Most anteriorly, the fused presphenoid cartilages become a vertical cartilaginous plate (*mesethmoid cartilage*) within the nasal septum. The mesethmoid cartilage ossifies at birth into the *perpendicular plate of the ethmoid bone*, its upper edge forming the *crista galli* that separates the olfactory bulbs (Fig. 8–3).

The capsules surrounding the nasal and otic (vestibulocochlear) sense organs chondrify and fuse to the cartilages of the cranial base. The *nasal capsule* (ectethmoid)* chondrifies in the second month pc to form a box of cartilage with a roof and lateral walls divided by a median cartilage septum (mesethmoid) (Fig. 8–4). Ossification centers in the lateral walls form the *lateral masses* (*labyrinths*) of the ethmoid bone and the *inferior nasal concha* bones.

The median nasal septum remains cartilaginous except posteroinferiorly, where, in the membrane on each side of the septum, intramembranous ossification centers form the initially paired *vomer* bone, its two halves uniting from below before birth but containing intervening nasal septal cartilage until puberty. The vomerine alae extend posteriorly over the basisphenoid, forming the roof of the nasopharynx—a distinctively human feature. Appositional bone growth postnatally on the posterosupe-

---

*The nasal capsule and nasal septum have been postulated to arise from prechordal (preotic) sclerotomes; the myotomes of these prechordal somitomeres give rise to the extrinsic eye muscles.

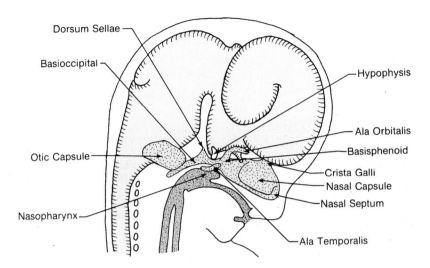

Figure 8–2 Schematic parasagittal section through the head of an early fetus (8th week). The chondrocranium cartilages are identified by blue colored stippling.

rior margins of the vomer contributes to nasal septal growth and indirectly to the downward and forward growth of the face.

The chondrified nasal capsules form the *cartilages of the nostrils* and the nasal septal cartilage. In the fetus, the septal cartilage intervenes between the cranial base above and the "premaxilla," vomer and palatal processes of the maxilla below.[†] The nasal septal cartilage is equivocally believed, by its growth, to play a role in the downward and forward growth of the midface (acting as a "functional matrix") (see p.173).

The *otic capsules* chondrify and fuse with the parachordal cartilages to ossify later as the mastoid and petrous portions of the temporal bones. The optic capsule does not chondrify in humans.

The initially separate centers of cranial base chondrification fuse into a single, irregular, and much-perforated *basal plate*. The early (prechondrification)

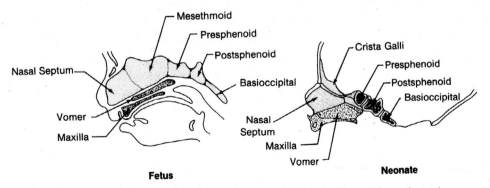

Figure 8–3 Cartilages of the fetal chondrocranium and their derivatives. (Blue color). The vomer and maxilla are of intramembranous origin. (Pink color).

---

[†]Postnatally, the septal cartilage may act as a strut to resist the compressive forces of incision, transferring these forces from the incisor region to the sphenoid bone.

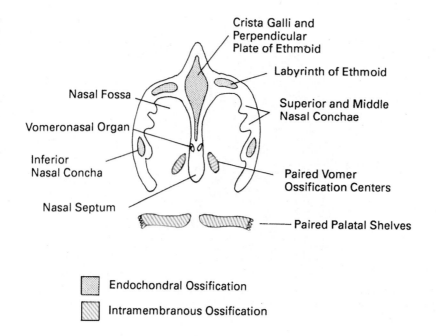

Figure 8–4 Schema of coronal section of nasal capsules and ossification centers.

establishment of the blood vessels, cranial nerves and spinal cord between the developing brain and its extracranial contacts is responsible for the numerous perforations (foramina) in the cartilage basal plate and the subsequent osseous cranial floor (see Fig. 8–1).

The ossifying chondrocranium meets the ossifying desmocranium to form the neurocranium. The developing brain lies in the shallow groove formed by the chondrocranium. The deep central hypophyseal fossa is bounded by the presphenoid cartilage of the tuberculum sellae anteriorly and the postsphenoid cartilage of the dorsum sellae posteriorly.

The fibers of the olfactory nerve (I) determine the perforations of the cribriform plate of the ethmoid bone. Extensions of the orbitosphenoid cartilage around the optic nerve (II) and ophthalmic artery, when fused with the cranial part of the basal plate, form the *optic foramen*. The space between the orbitosphenoid and alisphenoid cartilages is retained, a pathway for the oculomotor (III), trochlear (IV), ophthalmic ($V^1$) and abducens (VI) nerves and the ophthalmic veins, as the *superior orbital fissure*. The junction of the alisphenoid (greater wing) and presphenoid cartilages of the sphenoid bone is interrupted by pathways of the maxillary nerve ($V^2$) to create the *foramen rotundum*, and of the mandibular nerve ($V^3$) to create the *foramen ovale*. The middle meningeal artery forms the *foramen spinosum*. Persistence of the cartilage between the ossification sites of the alisphenoid and the otic capsule accounts for the *foramen lacerum*. Ossification around the internal carotid artery accounts for its canal, interposed at the junction of the alisphenoid and postsphenoid cartilages and otic capsule. The passage of the facial (VII) and vestibulocochlear (VIII) nerves through the otic capsule ensures patency of the *internal acoustic meatus*. The glossopharyngeal